70/-

KU-347-707

1 Morrell Cohen
2 Dick Levins
3 Jim Burns
4 Stuart Kauffman
5 Christopher Longuet-Higgins
6 Ruth Sager
7 Dick Lewontin
8 C. H. Waddington

1   2   3   4

5   6   7   8

Towards a Theoretical Biology

3. DRAFTS

Towards a Theoretical Biology

3. DRAFTS

an IUBS symposium
edited by C.H.Waddington

Edinburgh University Press

© 1970 International Union of
Biological Sciences & Edinburgh University Press
22 George Square, Edinburgh
85224 175 5
North America
Aldine Publishing Company
529 South Wabash Avenue, Chicago
First published 1970
Library of Congress
Catalog Card Number 68-19881
Printed in Great Britain by
R. & R. Clark Ltd., Edinburgh

Preface

This is the third volume of papers issuing from the Symposia on Theoretical Biology organized by the International Union of Biological Sciences at the Rockefeller Foundation's Villa Serbelloni. The emphasis has now become more sharply focussed. Most of the participants agree that the major problem which emerged from the earlier meetings is that of finding methods of dealing with the great complexity of biological systems. Molecular biology has given us considerable insight into the nature of the elementary units and processes of life, but to understand how these are put together to form systems which are usually too complicated to be analysed completely, but yet exhibit global properties of considerable simplicity, presents biologists with an intellectual challenge which the physico-chemical sciences scarcely find it necessary to face.

The problem is approached here from several different angles ; by Elsasser, Kerner, and Pattee from quantum physics, by René Thom from topology, by Garstens from statistical mechanics. A novel and very stimulating discussion is recorded in the papers grouped around Kauffman's remarkable demonstration that the behaviour of randomly constructed networks exhibits a surprising degree of orderly simplicity. Fraser, Lewontin, and Levins present some thoughtful analyses of complexities in such basic biological processes as the genetic control of differentiation, evolution, and ecology. Waddington attempts to answer the questions : what kinds of theories should we wish to have in connection with developmental biology ? and have we got them ? Wolpert, and Goodwin and Cohen carry on this theme by providing, for one of the most difficult of all problems, morphogenesis, a comprehensive theory on lines which will appear very unexpected to most biologists ; this is one of the most indubitably original contributions to be directly produced by these meetings. The book closes with a characteristically elegant and profound essay by Longuet-Higgins, cast in the form of a conversation between a physicist (his old self ?) and a biologist (his new one ?).

Again, we were very happy to be the guests of the Rockefeller Foundation at the Villa Serbelloni at Bellagio on Lake Como, and to be welcomed and made at home by Dr and Mrs John Marshall. To them, and to the International Union of Biological Sciences, I know I can express the heartfelt gratitude of all the participants.

C. H. WADDINGTON

Contents

Biological stability

B. C. Goodwin
University of Sussex

Analytical v. synthetic; or sensation v. intuition. Biological systems are engaged in a perpetual process of self-maintenance and self-realisation directed by internally-defined criteria of stability and organization. Phenomenologically it is this attribute of living beings which allows us to identify them as discrete, autonomous systems : autonomous not in the sense that they are independent of their environment, but in the sense that their 'goals' are different from those of the physical environment, and these goals are internally defined. That is to say, their states of stability, towards which they are constantly moving and away from which they are constantly being perturbed, are defined according to different criteria from the stable states of the natural physical world. These criteria are the laws of biological organization, the constraints which govern the process where-by the organism realises itself as a biological system. The aim of these symposia on theoretical biology has been to discover the essential nature of these laws, and to develop a language and a formalism which is 'appropriate' to the analysis of the living process.

As has emerged from the first two volumes, there is a considerable diversity of opinion regarding the appropriate language which should be used in biology ranging from that of physics and chemistry, through molecular biology, to auto-mata theory, analytical topology, and philosophy. There is an equivalent di-versity of opinion regarding the essence of the biological process. Furthermore, there emerged particularly clearly in volume 2, in the discussions between Bohm, Maynard Smith, Waddington, Grene, and Longuet-Higgins, a difference in taste regarding the degree of analytical rigour which a 'scientific' language ought to have. Maynard Smith and Longuet-Higgins clearly prefer clarity and nicety, and are willing to run the risk of taking too narrow a view for the sake of precision and fidelity to 'fact'. Bohm, Grene, and Waddington appear to feel that intel-lectual freedom and creative insight are more important at this stage in the development of biological concepts, and so prefer to conduct their discussion in a language much closer to the roots of experience, which necessarily borders on metaphysics and 'poetry'.

The distinction between these two viewpoints is a very fundamental one and reflects, I believe, fairly deeply opposed polarities of psychic state. These, in

1

turn, reflect alternative modes of coping with reality and ordering experience, which are usually combined in varying degrees in any individual. The particular polarity I have in mind is that which Jung (*Collected Works*, vol. 6) has described between Sensation and Intuition, two ends of one spectral axis of psychological types. Since most participants at the conference were fairly highly polarized in the same way along the other Jungian axis, that opposing Thought and Feeling, Thought being dominant, there was little tension arising from this potential opposition. This is perhaps fortunate, since enough heat was generated from collisions between those whose picture of the world derives largely from the logical ordering of their sensations, with an emphasis on what is referred to as 'facts', and those who wish to order their experience in terms of intuitive perception which both subsumes fact and goes beyond it, but sometimes appears to pay little attention to these elements of 'reality'. A beautiful and lucid analysis of these levels of cognition has been given by Suzanne Langer in the context of a study of language and art in her volume *Philosophy in a New Key*, and *Mind: A Study in Human Feeling*. We were particularly fortunate in having Dr Langer present at the last symposium.

I drew attention to this distinction of attitude and approach to the analysis of biological process in my contribution to volume 1. There I discussed this opposition in terms of the digital v. the analogue approach in relation to the description of the cycle of growth and division in cells. I expressed my own preference for the analogue-dynamical language, and the belief that a more generalised dynamics could deal with the discontinuous–logical aspects of biological systems, which at present is the domain of automata theory. I am still inclined to this viewpoint, as will emerge from the present essay, wherein I describe some developments in the construction of the requisite dynamical theory. However, it may well be that, as in the duality of wave (analogue)–particle (digital) theories of matter, it is more useful to retain an equivalent duality in the analysis of biological phenomena, and not to try too obstinately to force the emergence of a unified theory to cope with the two aspects of their behaviour. As Scott Fitzgerald has put it, the definition of an intelligent man is one who can hold two contradictory ideas in his mind simultaneously.

Richard Gregory's analysis [1] of the nature of the process of object-recognition in animals is exceedingly relevant to this problem. He distinguishes between a synthetic, analogue-type machine, and an analytic, digital-type one. The former works in real time and maps its environment by an analogue process, storing information in the form of models which simulate, in some form, objects

2

in the outside world. The latter work in terms of symbolic operations, requiring a dictionary to translate the natural environment, suitably digitalised, into its symbols and their relations, and must accordingly function in its own time, as does a digital computer. It is ideally adapted to communication, since it operates in terms of explicit symbols and rules. Any biological process which is primarily concerned with information transfer in terms of discrete symbols, such as the higher forms of consciousness and language, may be expected to have features of the analytical type and be most easily analysed in such terms. The process of heredity, regarded as the transmission of mutable symbols (genes) from one generation to the next, is also analysable in these terms. It is the natural domain of automata and recursive function theory.

On the other hand, those features of biological process which involve the relatively continuous mapping of one domain into another, with the occasional sharp discontinuity, such as the adaptive response of the organism to its environment, the unfolding of the developmental process, and the lower levels of psychological behaviour, may be more easily analysed in terms of the properties and functioning of a synthetic, analogue-type machine or dynamical system. The choice of language, of formalism, lies before the investigator, and the only rule for this choice is the general one that the good butcher does the least cutting. We are all butchers to some extent in our analyses, but if we find the natural joints and perceive the normal modes of operation of the system we are interested in, then we will be able the more successfully to apply the appropriate analysis. Inappropriate analyses are not necessarily wrong; they are simply clumsy, and may be misleading.

For example, I think that Christopher Longuet-Higgins [2] makes a somewhat misleading statement when he says: 'Not until we can interpret the DNA of a new species without actually growing an individual from it will we be able to claim a full understanding of epigenesis.' He is making a point here about the usefulness of the concept of program in understanding the nature of biological process, and particularly in distinguishing it from the dynamical processes which are studied in the physical sciences. The 'program' is a set of instructions which order or constrain the epigenetic process in a particular manner. If this kind of computer analogy is going to be used in a biological context, I believe that the utmost clarity is required to avoid confusion. First, the DNA 'program' does not function in the same way as a computer program, because it is not referred to directly when cells take 'decisions' about state changes. The crucial points in the epigenetic process are the switching points, where cells must 'decide' which

3

of the alternative possible pathways of differentiation to follow. In computer terms, the cell must at such a point obey some sub-routine which takes the form of a conditional instruction : if such and such conditions are satisfied, do such and such. The implication of the computer analogy is that the cell computes its own state, looks at the DNA program for further instructions, and then changes state accordingly.

This is not in fact what a cell does, although a formal analogy can be made between the biochemical behaviour of a cell and the operation of an automaton following a program. It may seem elementary to insist that all the operations of the automaton must at some point be interpreted in biochemical and physio-logical terms, when discussing such a process as epigenesis, but I have been somewhat dismayed at the amount of confusion that has arisen because of a failure of those using the computer analogy to illustrate the operation of al-gorithmic instructions at the biochemical level. At this level, what actually happens in a cell sounds like **a** rather different process from that described above. Basically, biochemical 'switching' can be analysed in terms of a (non-linear) interaction between the variables describing the state of a cell, the concentra-tions of biochemical species, for example. These interactions are all, in turn, analysable in terms of reactions of the following type. Let us suppose that a particular metabolite, Y, is formed from two precursor metabolite molecules, U and V, the reaction being catalysed by an enzyme, E_1. And suppose that a second metabolite, Z, is formed from the precursors V and W, E_2 being the enzyme catalysing this reaction. If U, V, and E_1 are present simultaneously in sufficient concentration, but either W or E_2 is absent, then Y will be formed. If V, W, and E_2 are present but either U or E_1 is absent, Z will be formed. These statements constitute the biochemical form of the conditional sub-routine. There is not in fact any independent computation of cell state, reference to DNA for instructions, and subsequent change of state in accordance with these instruc-tions. The way the DNA operates is by determining *in part* the instantaneous state of the cell, by taking part in the regulation of previous rates of synthesis of enzymes and other macromolecules, and by specifying their kinetic character-istics (such as allosteric properties). The cell makes a 'decision' on the basis of its instantaneous state, which immediately determines what biochemical re-actions can take place. The operations of computing instantaneous state, looking up instructions on the DNA for the correct subsequent state transition, and making this transition, are thus not a 'natural' representation of the process; there is a logical but not an operational equivalence. If one is going to use this

4

language, it is necessary to point out that the relevant algorithm for the decision, while ultimately specified in part by the DNA, actually operates at the moment of decision through enzymes, other macromolecules, and metabolites. The properties of the enzymes are in fact part of the molecular phenotype of the organism, the genotype specifying in large part its nature. We may say that decisions are taken by the phenotype, whose properties are largely determined by the genotype.

Having recognized that the genotype actually acts in the form of constraints on the state transitions allowed to the phenotype, we may once again ask the question : What is the most appropriate language for describing this kind of system ? The computer analogy focuses on the nature and origin of the constraints in the phenotype and insists correctly that the epigenetic process of a particular species, for example, is largely directed by these constraints. Let me emphasize at this point that the DNA does not totally specify the epigenetic or any other biological process. There are initial conditions, unspecified by the DNA which operate as constraints also. The DNA is designed to operate in a particular cellular environment, on which it is dependent. We would not expect to be able to mix in a test tube the DNA of a protozoon, for example, together with all its other molecular constitutents, and observe the DNA directing the construction of the protozoon. The sufficient conditions for a biological process are not to be found in any part of the system, but in its total organization. The origin of life is, of course, a different problem. The process was started as a discontinuity ; now it survives as a continuity.

Describing the operation of the genotype in terms of the constraints it imposes on the state transitions available to a biological system is to use the language of dynamics, which may be a more natural one for the description of certain processes than that of automata theory. It is this language which Pattee [3] has been using in his very interesting analysis of the problem of self-reproduction in biological systems. The particular type of constraint which he has introduced and which, he argues, distinguishes biological processes from those usually studied in classical dynamics, is the quantum mechanical non holonomic constraint. This is an interesting idea, and I think that in some sense the non-holonomic constraint, or something very similar to it, is the correct description of the type of constraint which the genotype imposes on the phenotype. It is by means of such constraints that the DNA program makes itself felt in phenotypic behaviour.

▶ *The nature of biological processes.* I stated initially that the organism is engaged in continuous self-realisation, of perpetuating a state whose nature and

5

Biological stability

stability are defined according to criteria which are distinctively and characteristically biological. This process, observed at the psychological level, has been
called individuation by Jung (*Collected Works,* vol. 9 I), totalization by Cooper
[4]. It is the unceasing activity of self-organization and self-completion according to internally-defined criteria of stability in the organism. Morphologically this
process is observed as the replacement of structures lost by accident, such as the
regeneration of limbs in urodeles or the healing of skin after wounding. Dynamically, it is the re-establishment of mean levels and normal temporal ordering of
physiological processes after a disturbance, such as the restoration of mean blood
cell counts and the normal diurnal rhythm in numbers after haemorrhage, or the
recovery of the normal initiation time of DNA replication in the bacterial growth
cycle after thymine starvation. In an embryological context, Waddington [5]
has used the word individuation to refer to the process whereby embryonic
fields tend to result in complete, spatially-ordered and functionally-integrated
structures. Here again it refers to the self-organizing properties of the system.

I believe that this concept of individuation expresses an essential characteristic
of the living process at any level of complexity. This is the intrinsic tendency of
living systems to undergo a complete, unitary set of activities, which constitutes
a totality and generates a pattern in time or in space, or more commonly both,
which is recognizably individual and complete. At elementary levels such as the
bacterium this individuation process is the ordered unfolding of the sequence of
events, spatially and temporally organized (constrained), which is known as the
cycle of growth and division. This process is stable in relation to its capacity to
continue the pursuit of the cycle to completion, resulting in the production of
two individual entities, the daughter cells, in the face of environmental perturbations. A truncated or incomplete cycle is a failure : the resulting entities, if
they are formed, will be in some sense deficient. The cycle must be complete in
order to be successful; it must be a *total* cycle.

The individuation process always involves two somewhat contradictory or
opposing tendencies : the maintenance of the system as a distinct, single unit;
and the internal differentiation of the system into partially autonomous subsystems or sub-processes. These internal differentiations are imposed by incompatible requirements which cannot be simultaneously satisfied. In the cell cycle,
for example, it is not possible for the act of cell separation to occur simultaneously with the replication of chromosomes which separate at that division.
Clearly, chromosome replication must be complete before daughter chromosomes
can be physically partitioned between daughter cells. Thus replication must

6

precede cell separation, imposing temporal differentiation on the system. Again, DNA and ribosomes cannot occupy the same space within a cell, so there is necessarily a 'nuclear zone' in bacteria where the DNA is located, and consequently spatial differentiation. Bacteria have relatively few incompatibilities of this kind, either temporal or spatial, and consequently they are particularly simple systems. The more complex the demands which a system must satisfy, the more likely are incompatible requirements to arise, and the more highly differentiated will it be spatially, in its morphology, and temporally, in its behaviour. Hence organisms tend to increase in complexity as their environments become more complex, demanding more varied behaviour.

▶ *The nature of biological stability.* In what sense is the stability of the biological process different from the stability of the physical systems whose behaviour is most familiar to us? The classical example of a stable dynamic system is the marble in the bowl. After a perturbation, providing it does not exceed certain bounds, the marble will return to its equilibrium position at the bottom of the bowl. Such a system is asymptotically stable in relation to a point. It is the prototype of the stable physical system.

There is another general class of stable system about whose behaviour, in quantitative terms, rather little is known. This consists of systems which are asymptotically stable in relation to a causally closed cycle of events. It is to this class of stable system that the living organism belongs. Consider one of the most basic processes which occur in organisms : the intracellular replication of DNA. This involves a causal cycle of events, the detailed elucidation of which established molecular biology as an analytical science : DNA→RNA→Protein→DNA. This may also be regarded as a short-hand representation of the process of virus or cell reproduction, although we must make sure that, in using such an abbreviated description of this process, we are not implying that DNA replication is all that system reproduction involves. The point is that this illustrates the cyclic nature of the self-reproduction process. At the very heart of biology there is, then, the concept of cycle ; and biological systems at the level of organization of the virus or the single cell realise themselves by undergoing a cycle whose result is the production of more of the same systems. This is their natural mode of behaviour.

Such a causal cycle does not necessarily involve a dynamic cycle, an oscillation. Phage reproduction is an example where the closed, causal cycle of events occurs at the molecular level, and there is no observable oscillation of state variables in the system with a period equal to the phage reproduction time. This

7

is because there is no system that maintains its integrity during the process. The phage coat gets left behind when the DNA is injected into the bacterium, which provides the environment for DNA self-replication. The phage makes its own enzymes for this process, so the causal cycle is as described above, but the temporal ordering of the events is not essential to the success of the process. If, for example, coat protein is present in the system (infected bacterium) before a particular phage DNA strand starts replicating, this will not disturb the self-replication process.

The moment one imposes the constraint that what reproduces itself is a system, distinct from its environment and maintaining its integrity throughout the cycle, such as a cell, then a dynamic oscillation in the state of the system becomes essential : the causal cycle becomes, dynamically, a limit cycle in the state space of the cell defined in terms of intensive variables. This results from the fact that such a system must now be temporally organized, differentiated : for example, DNA replication must be complete before cell separation occurs, in order that each daughter cell receives a complete chromosome. An oscillation is required to provide phase information for the relative timing of initiation of DNA replication and cell division [6].

A system which operates in terms of non-degenerate cycles of self-reproduction will evidently have properties which are rather different from the more familiar point-stable systems of the inanimate world. First, its state at any moment of time is not uniquely determined by the state of its environment. Since it is engaged in a cycle of activities, it will undergo state transitions autonomously : that is, it will exhibit properties of 'self-animation'. This is unlike classical thermodynamic systems, whose states at equilibrium are uniquely determined by environmental parameters. There will evidently be continuous, on-going activities in a system which maintains itself as a distinct entity throughout the reproduction process, and the basic dynamic characteristic of this activity is that it is rhythmic or oscillatory. We are led thus to the conclusion that oscillatory behaviour is the fundamental dynamic mode of living, self-reproducing systems, as we known them at and above the cellular level. The oscillation is not imposed by the environment ; nor is it incidental to the living process. It is central to its organization.

It is upon this dynamic foundation that organisms evolve. This does not imply that all aspects of organismic behaviour necessarily involve oscillatory or rhythmic activity, although many appear to do so, as has been amply elaborated by Iberall [7]. The more primitive organisms such as bacteria and the unicellulars exist almost exclusively in the cyclic mode of self-reproduction, excepting quiescent states such as spores. However, the more highly evolved metazoa show complex dynamic transients within the overall, necessary self-reproduction cycle, epigenesis being one of the most interesting of these. How this process may itself depend in a fundamental manner on cellular limit cycles dynamically similar to the primitive cycle of unicellular forms has been described in detail by Goodwin and Cohen [8] in terms of a periodic wave-propagation theory of the developmental process. The complexity of dynamical behaviour that can arise in the epigenetic process has, furthermore, been analysed brilliantly and described by Thom [9, 10] using the concepts of analytical topology. His theory of structural stability shows how instabilities can arise and unfold in a well-defined manner in a dynamical system when the instability is of the type which he classifies as an elementary catastrophe. Such controlled discontinuities may represent spatial and temporal differentiations of a particularly dramatic kind, which have the result of resolving certain incompatibilities, which arise as a result of the individuation processes of epigenesis. The occurrence of these discontinuities produces critical periods in this process, but their establishment and realisation contribute to the stability of the overall process.

▶ *The dynamics of self-reproduction.* I have stated that living organisms belong to a class of system whose stability characteristics are intrinsically different from those which apply to familiar physical systems; that they are stable in relation to a closed cycle of events, not in relation to a point; and that, when the organism is of a level of complexity equal to that of a cell or above, then this closed cycle of events has the characteristics of a limit cycle. The oscillation in the state of the system that necessarily occurs during the bacterial or other cell cycle has these characteristics because, after a perturbation, the oscillation returns to its original frequency and amplitude. Such perturbations can include relatively severe ones, such as spore formation, in which the oscillation goes 'dead'. When spores germinate, the oscillation starts again.

This general analysis suggests that there is something quite fundamental about the concept of a limit cycle in biological systems; that systems whose stability is characterized in relation to such a cycle have properties which distinguish them markedly from those with point stability. Some of these properties have been mentioned already. It would appear that many of the holistic concepts of biology arise from this fundamental but simple organizational basis. The primitive concept in biology of self, arising from the existence of continuous, on-going, 'purposeful' activity (stable in relation to a total cycle), may be analysed in terms of such self-sustaining cycles. The transition from life to death is the interruption of the cycle in such a manner that it cannot start again : a causal link has been destroyed, for example. Sporulation evidently involves the transition of the bacterial system to a state where it is both resistant to adverse conditions and able to restart the cycle when conditions are again favourable. It involves a quasi-death of the system.

Organization and stability in relation to a cycle is not by any means an exclusively biological characteristic. There are physical systems which show this kind of behaviour. One of the most familiar is the internal combusion engine. This system is not self-exciting : it needs an initial kick to start if off, a discontinuous or 'catastrophic' initiation. There is a well-ordered cycle of events which repeats with regular frequency for given environmental conditions : fuel and air intake, and so on. This cycle does not, of course, involve self-reproduction, and in this basic respect it differs from the biological cycle under consideration. One very obvious conclusion emerges immediately from these elementary considerations : the stable cyclic systems we are considering cannot be characterized and studied by the concepts and methods of equilibrium thermodynamics, which is concerned with the behaviour of a particular class of system with point stability. Autonomous cyclic operation can occur only in systems whose steady state is displaced from thermodynamic equilibrium. The detailed analysis of the properties of such systems initiated by such investigators as Prigogine and Glansdorff [11], and Lefever, Nicolis and Prigogine [12] promises a very rich harvest of new concepts in a field whose development has been slow but interesting, the thermodynamics of irreversible processes.

In characterizing the dynamic properties of biological self-reproduction in terms of a limit cycle, I have been considering primarily the topological and stability aspects of such a process. This characterization ignores nearly all the conceptual problems of self-reproduction such as have been considered by Pattee [3] and Arbib [13]. It incorporates only the stability properties of the process and the notion of cycle. In view of the basic nature of such processes in biology, however, it is of some interest to consider the problem of constructing a theory which can be applied to the quantitative analysis of systems organized dynamically on the basis of a limit cycle, or a set of interacting limit cycles. Such a theory might be expected to be of use in analysing the properties of this general class of system.

One way of proceeding with the construction of such a theory is to follow the conceptual approach of the physical sciences and ask the question : is something conserved in the motion of a dynamical system whose basic mode of activity is a limit cycle? If some quantity is conserved, then we might expect that it in some sense contains the dynamical essence of the system, in the same way that energy contains the essential properties of physical systems required to define their state. Now the important property of the energy invariant in physics is that what it 'contains' is the dynamics of the system; that is, the dynamical equations which describe the unfolding of the physical process can be obtained from the energy integral by well-defined procedures.

In the case of a limit cycle, the dynamical behaviour of the system consists not only of the actual limit cycle itself, but also of all the trajectories leading to the limit cycle, from inside and outside the cycle. Any conserved quantity, any constant of the motion for the system as a whole, must be defined on these trajectories and the trajectories must themselves be derivable from the conserved quantity. Now it has been established, by the theorem of Lie and König [14, 15], that such quantities, known as integral invariants, do exist for dynamical systems with limit cycles; in fact, they exist for any dynamical system which satisfies certain very general constraints on the continuity properties of the trajectories. However, in the general case it is not possible to derive a time-independent expression for the conserved quantity. The question of interest for our present enquiry is : can we derive such an expression for the case of limit cycles, which constitute a very specific class of dynamical system? The answer to this is yes, providing we transform the limit cycle into what we might call its 'canonical form' : a circle on a plane (*see* Goodwin, in [19]). Any limit cycle in any number of dimensions is tranformable to such a canonical form by a diffeomorphism; but again this is an existence theorem, and for an arbitrarily complex limit cycle the exact transformation would be a horrible one.

So for the time being we must restrict our attention to the behaviour of systems whose dynamical motion is that of a canonical limit cycle, or a set of such interacting cycles. This may not be such a restriction as it appears. Any arbitrary limit cycle may be decomposable to a set of interacting canonical limit cycles, in a similar way to that whereby any oscillation may be decomposed into a sum of sinusoidal oscillations by the Fourier expansion. The Fourier theorem is not applicable to non-linear systems since it depends upon the principle of superposition, which is not satisfied when there are non-linearities in the equations. The essence of non-linear behaviour in general emerges from interactions between variables. It may be possible to preserve this essence for the case of systems with limit cycles by a (non-unique) decomposition into a set of interacting canonical limit cycles. But at the moment this idea is highly speculative.

Biological stability

We may anticipate that an analysis of systems with limit cycle behaviour in terms of constants of the motion defined on their trajectories will have a rather different emphasis from that which emerges from the application of this procedure to the conservative systems of physics. The ideal or model system in physics is the one which conserves energy, and its prototype is the ideal pendulum which continues to oscillate indefinitely once set in motion. The stability properties of such a system are actually rather peculiar. They are those of the neutrally stable system, like the billiard ball on the flat billiard table : there is no preferred position of stability; the ball will stay wherever it is placed. The simple pendulum has, similarly, no preferred orbit; it will oscillate with whatever amplitude it is given initially. Each different amplitude gives it a different total energy, but none is preferred over any other.

The space of trajectories in which this kind of system moves is like the surface of a billiard table in the sense that it is 'flat' and smooth, with no attractor points or sets where the trajectories become dense. It is rather extraordinary that physics has been so successful in analysing the physical world in terms of such ideal systems, since there are no real systems which behave in this way. Physics received its basic structure from the analysis of planetary motion, in which the dissipative forces are in fact very small, ignorable. When physics shifted its attention from planets to molecules, it initially preserved the planetary ideal in the form of such constructions as the ideal gas : energy was still conserved, since the dissipative interactions were still very small.

Physics encountered its first non-ignorable stability problem in the form of the quantum of radiant energy and the adsorption spectra of the elements. Bohr [16] referred to the problem in terms of preferred orbits of electrons. This statement of the problem is precisely the one we are dealing with : how does one describe and analyse the behaviour of a population of systems which have attractor sets in phase space, such as limit cycles, towards which the trajectories of the systems move? The solution to this problem adopted by the physicists, after a number of false starts, was a stroke of remarkable ingenuity and good fortune : the discrete spectrum of stable states available to an atomic system could be put in 1–1 correspondence (almost) with the discrete spectrum of the solutions of a wave equation of a type already used in physics, the Schrödinger equation. This was partly an empirical discovery, partly a theoretical one : that is to say, much of the justification for the use of this equation came *ex post facto*. What appeared to be the final justification for the whole procedure came with the publication of von Neumann's treatise *The Mathematical Foundations of*

Quantum Mechanics in 1932. So complete was von Neumann's argument that it seemed to close forever further enquiry into the justification for the whole manner in which physics had apparently solved this very fundamental problem. The dogma arose that there could not be any analysis of the nature of the preferred states in atomic systems in terms of variables which underlay and accounted for these orbits as, for example, limit cycles in a dynamical space, because such variables could never be observed; they were hidden from experimental probing by the Heisenberg principle, hence forbidden in a scientific theory. In terms of such variables, the observed quantization of energy might be explained in terms of nested or concentric limit cycles of the type described by Poincaré. Only recently has this whole area of enquiry, which is an exceedingly interesting one, been re-opened by Bohm's demonstration of an incompleteness in von Neumann's axioms. We may now be about to witness a return in physics to this problem, which was so neatly side-stepped by the Schrödinger equation.

It is possible that something like this famous wave equation will be found for the description of the stable states of certain biological systems. This would depend upon the discovery of an analogue of the quantum of energy in biological behaviour, and there is as yet little evidence for this. However, even if appropriate quanta of action are found for biological behaviour, the physical-type procedure for analysing quantized systems would be inadequate. This is because in biological systems the variables which underlie observed behavioural states and which define the state space of the system are in fact observable. There is no Heisenberg principle which says that they are inaccessible to measurement. For example, a bacterium in a particular medium at defined temperature, O_2 tension, and so on, has a well-defined, unique mean generation time. This generation time is a 'macroscopic' observable, the resultant of the action and interaction of many observable microscopic or molecular variables, the concentrations of various molecular species in the system which are measurable and define its state at any one time. A theory of the cell cycle in bacteria must proceed from the molecular to the cellular level : from concentrations and their variations to generation times, which is the time for one complete circuit round the limit cycle. It is not sufficient to describe the system solely in terms of a quantum of behaviour, the complete cell cycle, although this is indeed a biological quantum of action. Such a quantum I referred to previously as a set of activities which is in some biological sense complete, total, realised, and is the basic feature of the individuation or the totalization process which characterizes biological acts. An adequate theory of the cell cycle, and of all biological process, requires that the observed quanta of behaviour be explained in terms of the dynamical behaviour of the observable variables which underlie and 'explain' the stability properties of the whole system.

The phase space of a biological system is not, then, a space evenly populated with phase points like that of physical systems which are described by classical Hamiltonians. It has attractor sets, preferred orbits, as in quantum mechanics. But we can observe the variables in this space; so we cannot avoid basing our dynamical theory of biological processes on the behaviour of trajectories in this space of 'microscopic' variables. If we wish to construct a theory in terms of constants of the motion for systems with asymptotic stability, then these constants must refer to some property of the system which is basically different from that which is contained in the energy integral. The new constant must refer to stability features : it is some property of stability that is conserved, not energy. Energy is in fact dissipated by these stable systems; and the more stable they are, the more rapidly do they dissipate energy. Thus stability is bought at the price of energy dissipation. A theory based upon stability properties must, therefore, be a non-equilibrium theory in the thermodynamic sense of the word, and we may expect that there will be some relationship between the conserved quantity, reflecting stability, and the rate of entropy production. It is to be anticipated that such a theory will look rather different from classical or quantum statistical thermodynamics, which are based upon the concept of energy conservation. A treatment of classical statistical mechanics in terms of stability theory has been given by Ulinski [17] in a very interesting report which illustrates clearly the particular kind of stability which these conservative systems have.

The essential difference between a neutrally stable system (one which is stable but not asymptotically stable; see reference [18]) and one with asymptotic stability is a topological one. A neutrally stable system is structurally unstable; that is, a small parametric perturbation of the system, such as adding a small dissipative term, profoundly alters the behaviour of the system. An asymptotically stable system, on the other hand, is structurally stable (for a definition of structural stability, see reference [10]). Thus from a topological point of view these are two quite different types of system. It has often been suggested that a biological theory should be based upon topological considerations, usually with the implication that the best we can hope for in biology is a qualitative theory of phenomena. The power of the topological approach is to be found in the generality of its analysis; it gives one both a high level of abstraction and analytical precision. This is amply illustrated in Thom's work. However, I believe that the qualitative insights provided by topology into the behaviour of dynamical systems must be combined with quantitative analysis in order to satisfy the demands of analytical biology. A theory of the cell cycle must do more than describe the nature of the stability of the system, for example. It must allow one to calculate means and variances of variables such as the generation time, cell size, frequencies of genes, concentrations of enzymes, and so on. One way of proceeding with this problem is to construct a statistical dynamics which is based upon a topological analysis of stability. By this I mean to identify the topological characteristics of the growing and dividing cell as a dynamical system; to derive a function which contains these dynamical characteristics, a constant of the motion; and then to use this function in the construction of a statistical mechanics. Such a theory would in fact be qualitative in so far as it is constructed upon a canonical representative of the dynamical class to which the system of interest belongs. It would also be quantitative since it would allow one to make calculations of quantities which describe the statistical properties of the ensemble obeying the canonical equations. A decomposition procedure for reducing an arbitrary limit cycle to a set of interacting canonical limit cycles would make the theory fully quantitative. A mathematical treatment of this approach to the analysis of the cell cycle is in preparation for publication [19].

▶ *Conclusions.* In the course of this essay I have shifted gradually from the use of relatively loose, somewhat emotive terms such as self-realisation, individuation, unitary, to the more precise terms of dynamics such as trajectory, limit cycle, stability, quantum. The translation of one set of terms into the other is fairly obvious. In so far as it is obvious, the language of dynamics is a natural one for describing and analysing biological process. However, this is by no means a sufficient reason for adopting this language. It may be appropriate for describing certain aspects of the process, but not others. What is in fact totally absent from the type of analysis I have given is the concept of constraint; in particular, the exact dynamical representation of the type of constraint which is imposed by the genotype of the organism. This is where the computer analogy exerts its attraction, since there is an analogy between the operation of the genome and the function of a program. I stated that the dynamical concept of the non-holonomic constraint, introduced by Pattee into the context of the study of self-reproducing systems, may well function as the correct translation of the ordering, constraining, or selecting influence exerted by the genes on the dynamics of the biological process. This influence makes itself felt through the parameters of the dynamical

15

process, and its effect is to reduce the number of actual trajectories of the system in relation to the virtual ones.

The exact topological representation of this kind of constraint is not very clear, and in this respect analytical topology does not yet provide a complete language for the description and analysis of the biological process. However, I may well be speaking out of ignorance at this point. The characterization of the properties of dynamical systems in terms of parametric variation, which is the domain of structural stability, a relatively new, rapidly developing, and intensely interesting area of analytical topology, may be expected to provide many insights into the nature of biological organization, and in particular the manner in which that characteristically biological constraining influence, the genome, exerts its influence on the development and functioning of the phenotype. This would indeed give general dynamics a complete and, I believe, appropriate language for the description and analysis of biological process, including both the analytical and the synthetic levels of operation. These two features of biological process must in fact operate in a single system, the total organism, and so be characterisable as aspects of one model system. Dynamics has been more preoccupied in the past with continuous motion in smooth spaces than with discontinuities in spaces with attractor sets. Hence its domain of utility has been a very restricted one since, as I have been arguing, the stability properties of most biological systems are simply not compatible with the properties of either linear or conservative dynamical systems. The necessary extensions of dynamics into the non-linear, non-conservative realm is slow but exceedingly interesting, and for the analysis of certain biological problems I believe that this is the only possible satisfactory procedure. Using theories designed to fit the behaviour of physical systems or machines will work if one is judicious in the choice of biological process which one chooses to analyse. But if biological systems have organizational criteria which are distinct from those of familiar physical systems, as I have been arguing, then it is unaviodable that a biological theory will have a different axiomatic foundation from physical theory. The nature of this difference is, of course, what has been debated at these symposia. I have argued that self-reproduction is a basic axiom for biological process, and that this has certain dynamical consequences of a very clear kind in systems whose complexity is that of the bacterial cell or greater. The full development of theory from these postulates will be a slow but interesting process.

B. C. Goodwin

References

1. R. Gregory, On how so little information controls so much behaviour. In (C. H. Waddington, ed.) *Towards a Theoretical Biology : 2, Sketches* p. 236 (Edinburgh University Press 1969).

2. C. Longuet-Higgins, What biology is about. In (C. H. Waddington, ed.) *Towards a Theoretical Biology : 2, Sketches* p. 227 (Edinburgh University Press 1969).

3. H. H. Pattee, The physical basis of coding and reliability in biological evolution. In (C. H. Waddington, ed.) *Towards a Theoretical Biology : 1, Prolegomena* (Edinburgh University Press 1968).

4. D. Cooper, *Psychiatry and Anti-Psychiatry* (London : Tavistock Publications 1967).

5. C. H. Waddington, *Principles of Embryology* (London : Allen and Unwin 1956).

6. B. C. Goodwin, Growth dynamics and synchronization of cells. *Symp. Soc. Gen. Microbiol. 19* (1969) 223.

7. A. Iberall, New thoughts on bio control. In (C. H. Waddington, ed.) *Towards a Theoretical Biology : 2, Sketches* p. 166 (Edinburgh University Press 1969).

8. B. C. Goodwin and M. H. Cohen, A phase-shift model for the spatial and temporal organization of developing systems. *J. Theoret. Biol. 25* (1969) 49.

9. R. Thom (1968), Une Théorie dynamique de la morphogenèse. In (C. H. Waddington, ed.) *Towards a Theoretical Biology : 1, Prolegomena* p. 152 (Edinburgh University Press 1968).

10. R. Thom, *Stabilité structurelle et morphogenèse* (W. A. Benjamin 1969).

11. I. Prigogine and P. Glansdorff, *Physica 31* (1965) 1242.

12. R. Lefever, G. Nicolis, and I. Prigogine, On the occurrence of oscillations around the steady state in systems of chemical reactions far from equilibrium. *J. Chem. Phys. 47* (1967) 1045.

13. M. Arbib, Self-reproducing automata – some implications for theoretical biology. In (C. H. Waddington, ed.) *Towards a Theoretical Biology : 2, Sketches* p. 204 (Edinburgh University Press 1969).

14. E. T. Whittaker, *Analytical Dynamics of Particles and Rigid Bodies* (Cambridge University Press 1937).

15. E. H. Kerner, Dynamical aspects of kinetics. *Bull. Math. Biophys. 26* (1964) 333.

16. N. Bohr, On the notions of causality and complementarity. *Dialectica 2* (1948) 312.

17. P. S. Ulinski and W. L. Kilmer, On relating the behaviour of a system to the behaviour of its constituents. *AFOSR-68-0217.* (Directorate of Information Sciences : Arlington, Virginia, 1968).

18. N. Minorsky, *Non-linear Oscillations* (D. van Nostrand 1962).

19. B. C. Goodwin. A model of the bacterial growth cycle : statistical dynamics of a system with asymptotic orbital stability. J. Theoret. Biol. (in press).

17

Behaviour of randomly constructed genetic nets : binary element nets

Stuart Kauffman
University of Cincinnati

A living thing is a complex net of interactions between thousands or millions of chemical species. A fundamental task of biology is to account for the origin and nature of metabolic stability in such systems in terms of the mechanisms which control biosynthesis. In the thermodynamics of gases, the mathematical laws of statistics bridge the gap between a chaos of colliding molecules and the simple order of the gas laws. In biology, a gene specifies a protein, and the output of one gene can control the rate of output of a second. The mathematical laws which engage large nets of interacting genes into biosynthetic coherence remain to be elucidated.

This article reports the behavior of large nets of randomly interconnected binary (on–off) 'genes'. The motives for this choice of model are many.

The analogy of genetic repression and derepression with digital computers has suggested to several authors [1–5] that the genome embodies complex switching circuits which constitute a program for metabolic stability, a cell differentiation.

It is a fundamental question whether metabolic stability and epigenesis require the genetic regulatory circuits to be precisely constructed. Has a fortunate evolutionary history selected only nets of highly ordered circuits which alone ensure metabolic stability; or are stability and epigenesis, even in nets of randomly interconnected regulatory circuits, to be expected as the probable consequence of as yet unknown mathematical laws? Are living things more akin to precisely programmed automata selected by evolution, or to randomly assembled automata whose characteristic behavior reflects their unorderly construction, no matter how evolution selected the surviving forms?

Evidence is presented below that large, randomly connected feedback nets of binary 'genes' behave with stability comparable to that in living things; that these systems undergo short stable cycles in the levels of their constituents; that the time course of these behavior cycles parallels and predicts the time required for cell replication in many phyla; that the number of distinguishable modes of behavior of one randomly constructed net predicts with considerable accuracy the number of cell types in an organism which embodies a genetic net of the same size; that, like cells, a randomly connected genetic net is capable of

18

differentiating directly from any one mode of behavior to, at most, a few of its other modes; and that these restricted transition possibilities between modes of behavior allow us to state a theory of differentiation which deduces the origin, sequence, branching, and cessation of differentiation as the expected behavior of randomly assembled genetic nets.

Several considerations suggest the advantage of modeling the gene as a binary device, able only to be on or off. The most fundamental of measures is the binary category scale. Use of these simplest devices facilitates study of the behavior of truly complex nets; the behavior of randomly connected, but then fixed, nets of binary components should provide a reliable guide to the behavior of similar systems whose components' behavior is described by continuous or probabilistic functions. Such a study is reported elsewhere in this book.

To study the behavior of randomly built nets requires definition of the population from which random sampling is to be done. A distinct advantage in the choice of a binary model for gene activity is that the number of different possible rules by which a finite number (K) of inputs may affect the output behavior of a binary element is finite — 2^{2^K}. (See Figure 1.) This allows construction of switching nets which are random in two different, but well defined, senses; the K inputs to each binary 'gene' may be chosen at random; the effect of those inputs on the recipient element's output behavior may be randomly decided by assigning at random to each element one of the possible 2^{2^K} Boolean functions of its inputs. Once built, the nets I have studied remained fixed in the choice of inputs to each gene, and their effect on its output.

The number of genes whose products directly affect the output of any gene is not, in general, known. Therefore, nets were studied in which each gene has direct inputs from all genes, nets with one input per gene, nets with two inputs per gene, and nets with three inputs per gene.

Since the autonomous, undriven behavior of a system must be elucidated before the effect of exogenous inputs can be understood, the behavior of switching nets free of external inputs has been studied.

The study of randomly constructed, but deterministic switching nets forms a poorly developed area of automata theory. Walker and Ashby [6] have examined the effect of the choice of Boolean function on the behavior of randomly inter-connected nets of binary elements. They simulated nets in which each of the 100 elements received a feedback input from itself, and randomly assigned inputs from two other elements. For each experiment, all elements of the net were assigned the same Boolean function.

19

Binary element nets

(a)

T			T+1
W	X	Y	Z
0	0	0	0
0	0	1	0
0	1	0	0
0	1	1	1
1	0	0	1
1	0	1	0
1	1	0	1
1	1	1	1

(b)

T		T+1
X	Y	Z
0	0	0
0	1	0
1	0	0
1	1	0

1

T		T+1
X	Y	Z
0	0	0
0	1	0
1	0	0
1	1	1

2

T		T+1
X	Y	Z
0	0	0
0	1	0
1	0	1
1	1	0

3

T		T+1
X	Y	Z
0	0	0
0	1	1
1	0	1
1	1	1

15

T		T+1
X	Y	Z
0	0	1
0	1	1
1	0	1
1	1	1

16

FIGURE 1
(a) *W, Xy* and *Y* are each binary devices which act as inputs to *Z*, another binary device. The 3 × 8 matrix of 1 and 0 below *W, X, Y* list the 8 possible configurations of input values to element *Z*. The columns under *Z* assigns to it the value it will assume one moment after each input configuration. (a) is one of the $2^{2^3} = 256$ Boolean functions of 3 variables. (b) The $2^{2^2} = 16$ Boolean functions of two input variables are derived by filling the column under *Z* with 1 and 0 in all possible (16) ways. Function 1 is *contradiction*, 2 is *and*, 16 is *tautology*.

These nets embody behavior cycles (described in detail below). Walker and Ashby found that the choice of the Boolean function assigned to all the elements markedly affected the length of these behavior cycles. Some functions (for example, 'and') yield very short cycles, others (for example, 'exclusive or') yield cycles of immense length.

Since there is no reason to suppose that, in living genetic reaction nets, all elements are assigned the same Boolean function I have studied nets in which all the 2^{2^K} possible Boolean functions are used, and assigned randomly, one to each element.

GENETIC MODEL
On these considerations, the gene is modeled as a binary device able to realize any one, but only one, of the possible Boolean functions of its *K* inputs. If the activity of a formal gene, for brevity, gene, at any time is 1, then the value of all its output lines at the time $T + 1$ is simultaneously 1. Thus, the state of the outputs of a gene at $T + 1$ depends on its activity at time *T* alone. For our logical

20

analysis, it is sufficient to allow time to occur in discrete, clocked moments :
$T = 1, 2, 3 \ldots$.

A *formal genetic net* is constructed by choosing a value of N, the number of elements comprising the net, and of K, the number of input lines to any gene. Each gene in the net receives exactly K inputs, one from each of K formal genes in N. Inputs arise only from members of N. On the average, each element has K output lines. Nets are randomly constructed in two distinct senses. The K inputs to each gene are chosen randomly; to each gene one of the 2^{2^K} Boolean functions of its K inputs is assigned randomly. After being assembled, these nets are deterministic. We assume that all genes compute one step in one clocked time unit.

Such a genetic net is a finite sequential automaton, a machine with a finite number of states and a function mapping each state into a subsequent state (see Figure 2). A state of the net is described by a row which lists the present value, 1 or 0, of each of the N elements of the net. Each gene can be independently on or off, thus there are just 2^N distinct states of a net of N binary elements.

If the system is placed in some state at time T, at $T + 1$ each gene scans the present value of each of its K inputs, consults its Boolean function, and assumes the value specified by the function for that input configuration. The net passes from a state to only one subsequent state; therefore, although two states may converge onto a single subsequent state, no state may diverge onto two subsequent states. (The system is state determined.)

There are a finite number of states. As the system passes along a sequence of states from any arbitrarily chosen initial state, it must eventually re-enter a state previously passed. Thereafter, the system cycles continuously through the re-entered set of states, called a *cycle*. The *cycle length* is defined as the number of states on a re-entrant cycle of behavior. Since more than one state may converge on a single state, the state re-entered need not be the arbitrarily chosen initial state.

A formal genetic net must contain at least one behavior cycle; it may contain more. The *number of cycles* embodied in a net is the number of different behavior cycles of which the net is capable. Since no state can diverge onto two subsequent states, no state on one cycle can simultaneously be on a second cycle. Different cycles in one net are behaviorally isolated from one another.

A *distance measure* comparing two states of the net may be defined as the number of genes with different values in the two states. (For example, the state

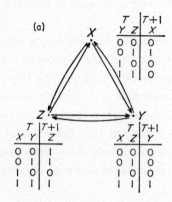

(a)

T Y Z	T+1 X
0 0	0
0 1	1
1 0	0
1 1	0

T X Y	T+1 Z
0 0	1
0 1	1
1 0	0
1 1	1

T X Z	T+1 Y
0 0	0
0 1	0
1 0	0
1 1	1

(b)

T X Y Z	T+1 X Y Z
0 0 0	0 0 1
0 0 1	1 0 1
0 1 0	0 0 1
0 1 1	0 0 1
1 0 0	0 0 0
1 0 1	1 1 0
1 1 0	0 0 1
1 1 1	0 1 1

(c)

```
                (010)
(100) ────→(001)────→(101)
              (011)       (110)
                ↑
              (111)
```

FIGURE 2

(a) A net of 3 binary elements, each of which receives inputs from the other two. The Boolean function assigned to each element is shown beside the element. (b) All possible states of the 3 element net are shown in the left 3×8 matrix below T. The subsequent state of the net at time T+1, shown in the matrix on the right, is derived from the inputs and functions shown in (a). (c) A kimatograph showing the sequence of state transitions leading into a state cycle of length 3. All states lie on one confluent. There are three run-ins to the single state cycle.

(00000) of a 5 gene net, and the state (00111) differ in the value of three elements.)

▶ *Noise*. As the net passes along a sequence of states on a cycle, one unit of noise may be introduced by arbitrarily changing the value of a single gene for one time moment. After perturbation, the system may return to the cycle perturbed, or run into a different cycle. In a net of size N, there are just N states which differ from any state in the value of just one gene. By perturbing all states on each cycle to all states a distance of one, a matrix may be obtained listing the total number of times the system returned to the cycle perturbed, or ran into any of the other possible cycles. Dividing the value in each cell of this matrix by its row total yields the corresponding matrix of transition probabilities between cycles, under the drive of random, one unit, noise. Such a matrix is a Markov chain. The probability of transition from one cycle to a second need not be identical with the probability of transition from the second to the first, Thus, state noise may induce asymmetric probabilities of transition between the independent behavior cycles of the net.

TOTALLY CONNECTED NETS, K=N

In random nets in which each element receives an input from all elements, the state subsequent to each state is chosen by sampling at random from an infinite supply of the 2^N distinct states of the net. The characteristics of such a random mapping of a finite set (2^N) of numbers into itself has been solved [7]. The expected length of the behavior cycle is the square root of the number (2^N) in the set. Therefore, in totally connected nets with 200 elements and 2^{200} states, the expected cycle length is $2^{100} \sim 10^{30}$ states. If the transition from one state to the next required one microsecond, then the time required for a net of 200 elements to traverse its cycle is about 10,000,000 times Hubbel's age of the universe. Totally connected, random nets are biologically impossible.

ONE CONNECTED NETS, K=1

Random nets in which each element receives just one input are no more biologically reasonable than totally connected nets. The structure of a one connected net breaks into separate loops of elements (as in Figure 2c with the direction of all arrows reversed). State cycles arise whose lengths are a maximum of two times the lowest common multiple of the set of structural loop lengths. For random nets as small as 200, the state cycles generally exceed several millions of the states in length [8]. One connected random nets possess behavior cycles capable of realization by no earthly organism.

TWO CONNECTED NETS, K=2

The behavior of randomly interconnected, deterministic nets in which each element received just two inputs from other elements is biologically reasonable. I have studied nets of 15, 50, 64, 100, 191, 1024, 4096 and 8191 elements, both by simulation on digital computers and analytically. Nets of 1000 elements possess $2^{1000} \sim 10^{300}$ possible states. The typical net is restricted to cycle among 12 of these states.

The program is used to construct a net of size N by random assignment of the two inputs and one of the $2^{2^2} = 16$ Boolean functions to each binary gene. The net is placed in an arbitrary initial state (for example, with each gene switched off) and, at successive time moments, computes its next state. Each of the sequence of states along a run-in is compared with all previous states, and when the present state is identical to a state of the system x moments previously, a cycle whose length is x states has been identified. If undisturbed, the system would cycle through these x states repeatedly.

▶ *Cycles.* Cycle lengths in such nets are exceptionally short. Data were obtained for at least 100 nets at each of several different sizes, and a histogram of the cycle lengths found in each size net was compiled. For each net size, the distribution of cycle lengths is markedly skewed toward short cycle lengths. Generally, the modal cycle length is less than the median length, which, in turn, is less than the mean cycle length. With $N = 400$ elements, the modal length is 2, the median is 8 and the mean is 98. Equilibrial states (those which successively become themselves) are common.

Among the 16 Boolean functions of two inputs, two are tautology and contradiction. An element assigned tautology is switched on regardless of the previous input values. An element assigned contradiction is constantly off. Thus, $2/16 = 1/8$ of the elements in a $K = 2$ random net are foci of constancy. These foci might be thought necessary to produce short behavior cycles. This is untrue. Nets were also studied in which these two functions were disallowed and the remaining 14 Boolean functions assigned equiprobably. The effect is to increase slightly the expected cycle length in nets of any given size and to shift the distribution of cycle lengths in nets of a given size from that found with all 16 Boolean functions. The distribution remains strongly skewed toward short cycle lengths, but the number of cycles of length one (equilibrial states) decreases. For example, with $N = 400$, the mode was 12, the median 32, and the mean 209. Deletion of tautology and contradiction increased the median cycle length in nets of 400 elements from 8 to 32 states.

Because the distribution of cycle lengths is highly skewed, the median cycle length seems the most representational length for nets of any size. The log of the median cycle length may be plotted against the log of the size of net, for nets with all 16 functions, and separately for nets without tautology and contradiction. The values in each condition appear non-linear in the log–log plot. The curves are initially steep, and flatten at larger values on N. In nets with tautology and contradiction allowed, the asymptotic log cycle length against log N is $\sim 0 \cdot 3$. In nets with tautology and contradiction disallowed, the asymptotic log cycle length is $\sim 0 \cdot 6$ log N. In this condition, the expected cycle length is just slightly greater than the square root of N ($0 \cdot 5$ in the log–log plot). As N increases, the median cycle length initially increases rapidly, then progressively more slowly. By projection, nets of 1,000,000 elements, with tautology and contradiction disallowed, possess behavior cycles of about 1,000 states in length — an extreme localisation of behavior among $2^{1,000,000}$ possible states.

▶ *Number of cycles.* The number of different state cycles — that is, the number of independent and different modes of behavior in these nets — are as surprisingly small as cycles are short.

By computer simulation, nets of 15, 50, 64, 100, 191, and 400 elements were studied. For each net, the system was placed successively in fifty arbitrarily chosen initial states, and the cycle discovered from each initial state was compared with previously discovered state cycles of that net. The median number of cycles per net is low; the distribution of the number of cycles per net around the median is skewed toward few cycles. Where $N = 400$, and neither tautology nor contradiction was allowed, the median number of cycles per net was 10. Presence or absence of tautology and contradiction does not seem to affect the number of cycles per net.

The log of the median number of cycles per net may be plotted against log N. The data appear to fall on a straight line with a slope of 0·5. Log number of cycles $\sim$0·5 log N. The expected number of modes of behavior is about $\sqrt{N}/2$. The number of cycles initially rises rapidly, then progressively slowly. By projection, nets of 1,000 elements will have about 16 cycles, and the nets of 1,000,000 about 500 modes of behavior.

Since only 50 run-ins to each net were made, the data probably underestimate the number of cycles per net. However, 200 run-ins per net rarely revealed more than 10 per cent more cycles than had the first 50 run-ins of the 200; the data, therefore, seem a good guide for the comparison of the number of cycles per net among nets of different sizes.

▶ *Distance between cycles.* The minimum difference which is possible between states on two cycles is 1 — a difference in the value of a single element. This distance occurs frequently, but the minimum distance may be as large as 0·3N. In nets of 100 elements, the median minimum distance between cycles is 5. The average distance between cycles is about 10. When a net embodies many cycles, these frequently form sets within which each cycle is a minimum distance of one from one to two members of the set. Between sets, the distance is larger, and may be as great as 0·3N.

▶ *Noise.* Perturbation, while running on a state cycle, has been studied in nets ranging from 15 to 2,000 elements, by reversing the value of each gene, releasing the net and observing if the net returned to the cycle perturbed. Nets larger than 400 elements used all 16 Boolean functions. In those less than 400, both conditions – with and without tautology and contradiction – were simulated. In general, the probability that a net returns to the cycle perturbed after arbitrary slight perturbation is between 0·85 and 0·95. Behavior in randomly connected binary nets is highly stable to infrequent noise.

One might have supposed that infrequent noise would induce a shift from each cycle to all others. This proves untrue. Transitions from a cycle are highly restricted; generally each cycle can shift to only one to six other cycles with probabilities of 0·01 to 0·05, and to a few others with probabilities between 0·01 and 0·0001. Most cycles cannot be reached directly from any other cycle.

Despite the restricted transition possibilities from each cycle, in many instances, the entire cycle set forms one ergodic region, that is there exists some path from each cycle to every cycle and back. Equally frequently, a subset of the cycles forms one ergodic region and the remaining cycles are transient cycles leading into the ergodic region, but not reachable from it. In the latter case, under infrequent noise, the system may progressively restrict the locale of its activity to the ergodic subset of cycles.

In no case when all possible single units of state noise were explored has more than one ergodic region been found. Restriction of perturbation to the first $0·6N$ of the N genes, however, has on one occasion yielded two ergodic regions. Further restriction of perturbation to $0·05N$ renders multiple ergodic sets probable.
▶ $K = 3$ nets. The occurrence of short cycle lengths and few cycles in random nets seems not to depend narrowly on an interconnection of two inputs per gene. I have simulated nets of 15, 20, 25, and 50 elements, each receiving three inputs from other elements, and allowed use of all $2^{2^3} = 256$ Boolean functions of three variables. Cycles were slightly longer, the number of cycles being about the same as comparable nets of connectivity two. These characteristic behaviors of randomly contructed nets seem to require only low connectivity to occur. The rate of their failure as K approaches N will require careful delineation.

DISCUSSION

It is surprising that randomly constructed nets, in which each element is directly affected by two others, embody short, stable behavior cycles. The immense restriction of behavior in a $K = 2$ net of 1,000 elements, limited to cycles a few hundred states in length, can only be appreciated in contrast to an expected state cycle length of 10^{150}. In a totally connected ($K = N$) net of the same size

10^{150} assumes its appropriate proportion when one remembers that 10^{23} estimates the age of the universe in microseconds.

Schrödinger [9] noted that high molecular specificity, guaranteed by quantum stabilization, is required for the precision of biosynthesis in living things. The behavior of these randomly connected nets discloses an unsuspected and, I believe, fundamental corollary to that precision. A molecular reaction net of high specificity *is* a net of low connectivity. To the extent that restricted behavior in binary element nets generalizes to continuous or probabilistic element nets of greater verisimilitude (see following chapter), the results suggest that high molecular specificity may be necessary both for precision of product formation, and to yield a system whose global chemical oscillatory behavior is brief and stable. Further, it would appear that nearly any biochemical net of high specificity would show highly localized behavior. . . . Evolution's task may have been less difficult than we thought.

The hypothesis that living genetic nets are randomly assembled does not imply that one gene of these nets lacks a specific effect on a second. It asserts that, if the 'wiring diagram' of the specific repression and derepression connections between genes were known, it would be topologically indistinguishable from a 'wiring diagram' generated by random assignment of specific interactions between genes. The hypothesis is consistent with both the random modifications of protein structure induced by mutation, and the lack of steric similarity between the molecule mediating end-product inhibition of an enzyme, and the substrate of that enzyme.

Biologically reasonable behavior in random nets occurs only if each element is directly affected by about the same low number of other elements as are macro-molecules in living things. This correspondence lends support to the hypothesis that living metabolic nets are randomly constructed.

CELL REPLICATION TIME

Among the most characteristic cyclic phenomena in cells is their replication. Van't Hof and Sparrow [10] have studied the minimum replication time in cells of several species of higher plants. They show the minimum cell replication time as a function of the DNA content per cell nucleus in six species of plants. The data fall nearly on a straight line. The authors conclude that, in higher organisms, minimum cell replication time is a linear function of the DNA content per nucleus.

Projection of this linear function predicts that cells without DNA will require several hours to replicate; bacteria with little DNA per cell require about 30 minutes to replicate. A curve of replication time from organisms with little DNA per cell to higher organisms must start near the origin, rise rapidly as the amount of DNA per cell increases, then rise more slowly as the DNA per cell continues to increase. Van't Hof and Sparrow [10] suggest the assumption of a second mechanism to control the time required for cell replication which would provide a steep linear slope from the origin, and intersect their observed linear function among higher plants. Choice of control mechanism would depend upon the nuclear content of DNA.

A single different principle, the hypothesis that living things are typical, randomly interconnected reaction nets, may be able to predict cell replication time as a function of the number of genes per cell throughout a wide range of phyla.

Estimates of the time required to switch a gene on or off lie between 5 and 90 seconds [11]. I will assume that about one minute suffices for a state transition in a real genetic net. Thus, if the model predicts a state cycle length of 100, the biochemical realization of the model should require about 100 minutes to traverse its cycle of oscillatory chemical concentrations.

In figure 3 I have plotted the logarithm of cell replication time in minutes against the logarithm of the estimated number of genes in that cell, for several species. The data include bacteria, protozoa, yeast, *Aspergillus*, sea urchin, chicken, mouse, rat, man, rabbit, dog, frog, and minimum cell replication time for *Vicia faba*, and several other plants. The number of genes per cell was estimated by comparison of its DNA per cell with that of *E. coli*, which Watson [12] has estimated to have about 2,000 genes. Based on these procedures, human cells embody about 2,000,000 genes.

The median cellular replication time for bacteria*, protozoa, chicken, mouse, and man are also shown in figure 3. It is apparent that these median replication times fall very nearly on a straight line whose slope on a log–log plot is 0·5. The expected replication time in minutes is therefore about the square root of the estimated number of genes. The square root of N increases rapidly with the initial rise in N, then more slowly as N continues to increase.

---

*Bacteria were assumed to have about the same DNA per cell content and to code for about 2,000 genes. In protozoa, the number of genes per cell is difficult to estimate due to the macronucleus. I have treated all protozoa as having about the same number of genes per cell, and estimated this number by dividing the cellular DNA content in *Tetrahymena* by the ratio of macronucleus DNA to micronucleus DNA in *Paramecium*.

I assume the DNA per cell in *Aspergillus nidulans* is about equal to that in *Neurospora crassa*. Rosenberger and Kessel chose growth media to yield disparate replication times in *Aspergillus* (1·4, 1·8, 3·7, 4·7, 7·0, 9·0 hr). I assume the first three represent relatively normal values.

TABLE I

Organism	DNA per Cell	Cell Replication Time
Bacteria	Watson [12]	Altman & Dittmer [13]
Protozoa	Nanney & Rudzinska [14]	Altman & Dittmer [13]
Sea Urchin	Sparrow & Evans [15]	Mazia [16]
Chicken	Vendrely [17]	Cleaver [18]
Mouse	Vendrely [17]	Cleaver [18]
Rat	Vendrely [17]	Cleaver [18]
Man	Vendrely [17]	Cleaver [18]
Rabbit	Vendrely [17]	Cleaver [18]
Dog	Vendrely [17]	Cleaver [18]
Frog	Vendrely [17]	Cleaver [18]
Vicia faba	Van't Hof & Sparrow [10]	Van't Hof & Sparrow [10]
Pisum sativum	Van't Hof & Sparrow [10]	Van't Hof & Sparrow [10]
Tradescantia paludosa	Van't Hof & Sparrow [10]	Van't Hof & Sparrow [10]
Tulipa kaufmanniana	Van't Hof & Sparrow [10]	Van't Hof & Sparrow [10]
Helianthus annuus	Van't Hof & Sparrow [10]	Van't Hof & Sparrow [10]
Trillium erectum	Van't Hof & Sparrow [10]	Van't Hof & Sparrow [10]
Aspergillus nidulans	Horowitz & Metzenberg [19]	Rosenberger & Kessel [20]
Saccharomyces cervesiae	Horowitz & Metzenberg [19]	Williamson [21]

The behavior of randomly interconnected binary element reaction nets predicts this observed relation between DNA content and replication time. The length of state cycles in random nets increases at almost the same rate as does cell replication times as the number of genes increases. Based on the assumption that a state transition requires about one minute, the model predicts a cycle time of about 50 minutes in a net of 2,000 elements, and 16 hours in a net of 1,000,000. The rate of increase of cycle lengths in nets with and without tautology and contradiction as N increases are shown superimposed on the biologic data of figure 3. Cell replication time falls between the two.

In the range of DNA per cell where Van't Hof and Sparrow [10] describe a linear relation between the DNA per cell and the minimum replication time, the relation between net size and cycle length in nets using all 15 Boolean functions is very nearly linear and of similar slope.

The model also appears to predict the distribution of replication times in cells with the same number of genes. Bacteria, with about the same number of genes − 2,000 − concentrate their replication times between 12 and 100 minutes, and scatter them up to 2,000 rarely. Random nets of 1,000 elements, using neither tautology nor contradiction, concentrate their state cycle lengths between 10 and 100 states, and scatter them up to 2,000 − 10,000 rarely. In figure 3 are several state cycle lengths in nets of 1,000 elements. The distribution is similar to that for bacterial replication times. Both distributions are skewed toward short

Binary element nets

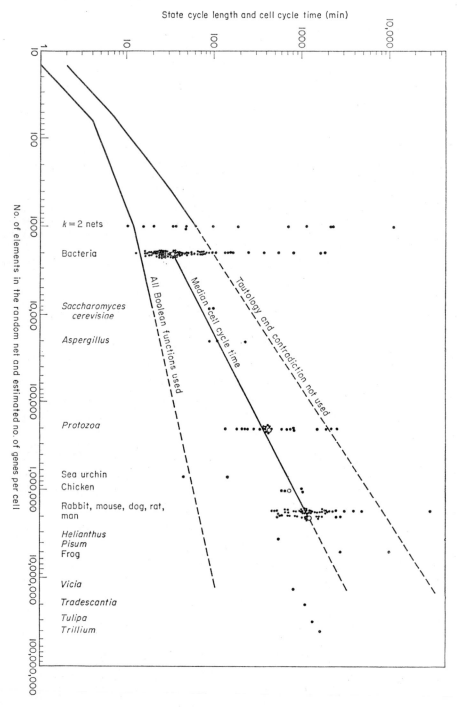

State cycle length and cell cycle time (min)

No. of elements in the random net and estimated no. of genes per cell

cycle lengths in a linear plot. A more rigorous test of their similarity lies in the fact that both remain skewed toward short cycles in a logarithmic plot.

Is this correspondence coincidental? Replication of the DNA in higher organisms is known to be initiated at many independent sites. Initiation of replication along any small segment of a chromosome is thought to require the activity of a 'replicon', and protein synthesis [16]. If these replicons form elements in the total metabolic net of the cell, depending for their own initiation upon the previous synthesis of other materials, it would not be unduly surprising that the periodicity of their activity, the S period, is bound by the periodicity of the entire metabolic net.

Although the behavior of binary element nets does predict cell replication times reasonably well, the correspondence must be viewed with caution. The real cell's genes presumably are neither binary nor synchronous devices. The study of more realistic stochastic nets (p. 38) suggests that obtaining well ordered cyclic sequences of oscillations of gene activity is not common. The mechanisms which control the sequencing of replicon activity in Eukaryotes are not yet understood.

CELLULAR DIFFERENTIATION

The principles underlying cellular differentiation remain among the most enigmatic in biology. We are required to explain the spontaneous generation of a multiplicity of cell types from the single zygote, to deduce a natural tendency of a system to become increasingly heterogeneous, then to stop differentiating.

Among the important characteristics of cell differentiation are: initiation of change; stabilization of change after cessation of its stimulus; the efficiency of biochemical noise in an inductive stimulus; a limit of five or six as the number of cell types which may differentiate directly from any cell type; progressive limitation in the number of developmental pathways open to any small region of the embryo; restricted periods during which a cell is competent to respond to an inductive stimulus; the indiscreteness of cell types, that is, the mutually

---

FIGURE 3

Logarithm of cell replication time in minutes plotted against the logarithm of the estimated number of genes per cell for various single cell organisms, and various cell types in several metazoan organisms. The data for the plants, *Vicia faba, Pisum*, etc. are the *minimum* replication times described by Van't Hof and Sparrow [10]. The solid line through the biologic data connect the median replication times of bacteria, protozoa, chicken, mouse and dog and rabbit, and man. Data from binary nets of 1,024 elements using neither tautology nor contradiction are included for comparison. Median cycle lengths in binary nets with and without tautology and contradiction, as a function of the number of elements in the net, are superimposed on the biologic data. Scale: 2×10^6 genes $= 6 \times 10^{-12}$ g. DNA per cell.

Binary element nets

exclusive constellations of properties by which cells differ; a requirement for a minimal and preferably heterogeneous cell mass to initiate differentiation in many instances, and to maintain it in some; the occurrence of metaplasia between undifferentiated cell types, or from an undifferentiated type to a specialized type, but the lack of metaplasia (the isolation) between specialized cell types; and the cessation of differentiation [22].

I believe many aspects of differentiation may be deducible from the typical behavior of randomly built genetic nets.

Cells are thought to differ due to differential expression of, rather than structural loss of, the genes. Differential activity of the genes raises at least two questions which are not always carefully distinguished : the capacity of the genome to behave in more than one mode; and mechanisms which insure the appropriate assignment of these modes to the proper cells. The second presumes the first.

Randomly assembled nets of binary elements behave in a multiplicity of distinct modes. Different state cycles embodied in a net are isolated from each other, for no state may be on two cycles. Thus, a multiplicity of state cycles, each a different temporal sequence of genetic activity, is to be expected in these randomly constructed genetic nets. It seems reasonable to identify one cell type with one state cycle. To the extent that this binary model, in which the expression of the 'gene' is potentially reversible at each clocked moment, is accurate, it demonstrates the common occurrence of multiple modes of behavior in a genetic system.

If this identification is reasonable, the typical number of cycles in a random 'genetic' net must be of the same order of magnitude as the number of cell types in organisms with the same number of genes.

Estimates of the number of cell types in an organism are hazardous, but the number in man may be placed at about 100; in annelid worms, at 57; in jellyfish, between 20 and 30; in hydra, between 11 and 17; in sponges, about 12–14; in *Neuspora crassa*, 5; in algae, 5; and in bacteria, 2; vegetative and spore, 3. The logarithms of the values are plotted against logarithms of the numbers of genes per cell in each organism, in figure 4. A straight line has been drawn through these values ; its slope is 0·5.

The logarithm of the number of independent cycles in a random net is also about 0·5 logarithm of the number of genes. By projection of these data, nets with about 16,000 genes (comparable to the sponges) should have about 120 cell types, and man, with an assumed 2,000,000 genes, about 700 cell types. These theoretical predictions are also plotted in figure 4.

32

The rate of increase in the number of cycles in random nets as N increases appears almost identical to the rate of increase in the number of cell types of an organism as the number of genes increases. The theoretical curve is shifted to the left, however, and predicts more cell types than are actually counted. The predictions remain well within an order of magnitude of the biologic data.

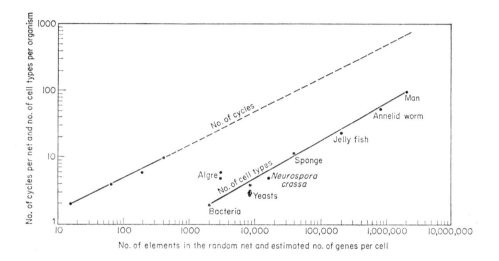

FIGURE 4
The logarithm of the number of cell types is plotted against the logarithm of the estimated number of genes per cell, and the logarithm of the median number of state cycles is plotted against logarithm N. The observed and theoretical slopes are about 0·5.

Caution is required for several reasons : large nets have not been simulated ; these nets use binary elements, nets of greater verisimilitude must be studied ; estimates of the number of cell types in an organism or the number of genes in that organism are only approximate.

Cells differ from one another in the possession of a constellation of properties which do not intergrade. Similarly, cycles in random nets may be compared for the minimum dissimilarity between their states. In nets of 100 elements, the minimum distance between cycles is commonly 0·05N to 0·25N. Like cell types, behavior cycles are generally separated from one another by a constellation of properties. Since no state may lead to two states, hence be on two cycles, state cycles, like cell types, are mutually exclusive.

If the multiplicity of modes of behavior in a random net helps elucidate the capacity of the genome to behave in more than one way, the appropriate

33

segregation of these modes of behavior to the correct cells requires explanation. Biochemical noise may play a very large role in directing the segregation. A theory which assigns to biochemical noise the task of segregation of different modes of genetic behavior to different cells offers great advantages. Biochemical noise is ubiquitous, unavoidable, and therefore in a sense reliable. It remains to show that biochemical noise in a randomly cross-coupled genetic net can produce orderly sequential segregation of behavior modes to the appropriate cells.

Perturbation of nets, behaving on cycles, by one unit of noise, generally had only a transient effect on the system's behavior. With a probability of about 0·9, the system returned to the cycle perturbed. The remaining $0·1N$ noisy inputs caused the system to shift from any cycle to at most one to six other cycles with probabilities greater than 0·01, and a few more with probabilities below 0·01. It is therefore of considerable interest that, throughout phylogeny, no cell seems to differentiate directly into more than a few other cell types. Restriction in the possible transitions between modes of behavior appears to be characteristic of both random nets and cell types.

The spontaneous generation of a multiplicity of cell types from a single cell type follows explicitly from this model. The occurrence of infrequent noise induces on the cycles of a randomly constructed net the transition probabilities between them which form a Markov Chain. Such a chain *must* have at least one ergodic region — a set of cycles each of which can reach all cycles of that set, but no other cycles. It *may* have transient cycles lying outside the single ergodic set, reaching into, but not reachable from, that set. It also *may* have more than a single ergodic set; each ergodic set must be isolated from all other ergodic sets, however all may be reachable from some single transient either directly, or via other transients. Let the net embody only a single ergodic set of cycles. If placed on any cycle in the set and perturbed by noise, the system will 'spontaneously' pass from cycle to cycle along the allowed transition pathways. An isolated cell would appear to oscillate among its modes of behavior, driven by external noise. If the net is a replicating cell, the clone will explore the permitted transition pathways between cycles and populate the ergodic set according to the asymptotic transition probabilities between the cycles. Cells will spontaneously start to change, pass down restricted pathways of development, populate the complete set of possible modes of behavior, and settle to some stable distribution. Differentiation will start, proceed, and stop at a stable distribution of cell types.

Since the net, by hypothesis, embodies only a single ergodic set, each cycle may reach all; differentiation of any cell cannot be stable, a cell of one type

34

should occasionally 'spontaneously' change to become a cell of a different type. Since all types may be reached from any type, deletion of a subset of cell types should create a net movement of cells into the types of cells removed – regeneration should occur.

Stable, irreversible differentiation would require either location in a microenvironment, in which the noise was sharply biased by the neighboring cell types, or, more fundamentally, a multiplicity of ergodic regions all reachable from at least one transient. Such a system corresponds to Waddington's epigenetic landscape, and is obtainable if 'noise' is limited, that is, if not all gene's values may be altered from outside of the net. In this case, the zygote would correspond to the Markov transient which reached into all ergodic regions. Clone members would pass down restricted pathways into various ergodic regions. Competence of limited duration would correspond to presence at a branch point between two ergodic regions. Progressive limitation of possible paths of differentiation would result from passing these branch points. Due to a multiplicity of ergodic regions, total regeneration, from remnants of less than all the ergodic sets, would not occur. Granting a birth and death rate for each cell type, the system would settle to a steady state of distribution of cell types.

Study of the typical behavior of randomly assembled determinate nets has barely begun. Further research is now needed to extend these results to larger nets; to study the effect of different numbers of inputs per element; to establish firmly the behavior of nets whose elements realize biologically appropriate continuous or probabilistic functions of their inputs; to find the effect of increasing levels of state noise, and more particularly of 'biased' noise to spatial proximity with other copies of the net behaving on different cycles, and to study the effect of mutation – random alteration in the structure of the net – on its behavior.

While the model has been developed to study cellular control processes, it is formally identical to nerve net models and may find application in other branches of science.

CONCLUSION

A living thing is a richly interconnected net of chemical reactions. One can little doubt that the earliest proto-organisms aggregated their reaction nets at random in the primeval seas; or that mutation continues to modify living metabolic nets in random ways.

It is a fundamental question whether two billion years of survival pressure have succeeded in selecting from a myriad of unorderly reaction nets those few

Binary element nets

improbable, that is, non-random and ordered, metabolic nets which alone behave with the stability requisite for life; or whether living things are akin to randomly constructed automata whose characteristic behavior reflects their unorderly construction no matter how evolution selected the surviving forms.

The data I have presented suggest : that large, randomly interconnected feedback nets of binary 'genes' behave with the stability requisite for life; that they undergo short stable cycles in the levels of their constituents; that the time required for these behavior cycles parallels and predicts the time required for cell replication in many phyla; that the number of distinguishable modes of behavior of a randomly constructed net predicts with considerable accuracy the number of cell types in an organism which embodies a genetic net of the same size; that, like cells, a random net is capable of differentiating directly from any one mode of behavior to, at most, a few of its other modes; and that these restricted transition possibilities between modes of behavior allow us to state a theory of differentiation which deduces the origin, sequence, branching, and cessation of differentiation as the expected behavior of randomly assembled reaction nets.

If original proto-organisms built their reaction nets randomly, it behoves the biologist to build an adequate theory of the behavior of these systems ; such a theory should elucidate the problems of biosynthetic organization faced by early living forms. But, if extant biota are also randomly constructed, then an adequate theory of the behavior of randomly assembled reaction nets may constitute an appropriate theory in which to describe and explain the metabolic behavior of nets throughout phylogeny.

References

1. F.Jacob and J.Monod, 21st Symposium Soc. Study of Development & Growth (Academic Press: London 1963).
2. M.J.Apter, *Cybernetics and Development* (Pergamon Press: Oxford 1966).
3. J.F,Bonner, *The Molecular Biology of Development* (Oxford University Press 1965).
4. M.Sugita, *J. Theoret. Biol. 4* (1963) 179.
5. S.A. Kauffman, *J. Theoret. Biol. 17* (1967) 483.
6. C.C.Walker and W.R.Ashby, *Kybernetic 3* (1965) 100.
7. H.Rubinand R. Sitgreave, *Probability Distributions Related to Random*
Transformations on a Finite Set (Tech. Report No. 19a, Appl. Maths. and Stats. Lab. Stanford University, 1954).
8. N.J.H.Slone, *Lengths of Cycle Time in Random Neural Networks* (Cornell University Press: Ithaca 1967).
9. E.Schrödinger, *What is Life ?* (Cambridge University Press 1944).
10. J.Van't Hof and A.H.Sparrow, *Proc. natn. Acad. Sci. USA 49* (1963) 897.
11. B.C.Goodwin, *Temporal Organization in Cells* (Academic Press: London 1963).
12. J.D.Watson, *Molecular Biology of the Gene* (W.A.Benjamin, Inc.: New York 1965).

13. P. L. Altman and D. S. Dittmer, *Growth*. (Fed. Am. Soc. exp. Biol.: Washington, D.C., 1962).

14. D. L. Nanney and M. A. Rudinska, Protozoa. In *The Cell*, vol. 4 (Academic Press: London 1960).

15. A. H. Sparrow and H. J. Evans, *Brookhaven Symposia in Biology, 14* (1961) 76 (Brookhaven Natl. Lab., Upton, N.Y.).

16. D. Mazia, Mitosis. In *The Cell*, vol. 3. (Academic Press: London 1961).

17. R. Vendrely, in *The Nucleic Acids*, vol. 2, p. 155 (Academic Press: New York 1955).

18. J. E. Cleaver, *Thymidine Metabolism and Cell Kinetics* (North Holland Press: Amsterdam 1967).

19. N. H. Horowitz and R. L. Metzenberg, *Ann. Rev. Biochem. 34* (1965) 527.

20. R. F. Rosenberger and M. Kessel, *J. Bacteriology 94* (1967) 1464.

21. D. H. Williamson (E. Zeuthen, ed.) *Synchrony in Cell Division and Growth* 351 (John Wiley & Sons: New York 1964).

22. C. Grobstein, Differentiation of Vertebrate Cells. In *The Cell*, vol. 1 (Academic Press: London 1959).

Behaviour of randomly constructed genetic nets : continuous element nets

Stuart Kauffman
University of Cincinnati

In the previous chapter, the gene was modeled as a binary device, and the behavior of large, randomly constructed nets of 'genes' was examined. The results indicated that, if the number of inputs per component was very low — two or three — then these nets behave in very orderly and restricted ways, show homeostasis by returning after perturbation, and can behave in only a few distinct orderly ways, each interpretable as a different 'cell type'.

In view of these promising beginnings, it is of considerable interest to know whether these 'good' behaviors of randomly built nets with two inputs per element are limited to binary element nets, or generalize to nets of more biologically realistic elements. The gene, in one sense, is a binary device. At any moment it either does or does not make an mRNA molecule. But, over a somewhat longer interval, presumably its rate of transcription can vary between some minimum and maximum, either as an 'N'ary or continuous device. Do randomly constructed nets of such Nary or continuous components show restricted and orderly behavior? In particular, do nets with two inputs per component behave in 'good' ways?

In an initial attempt to answer this question, it seemed reasonable to consider the simplest generalizations of binary deterministic nets. In reality, at least the following factors must be considered : A gene presumably transcribes an mRNA over some period — can it transcribe several simultaneously, at various stages of completion ? What are the half lives of different mRNAs ? Can mRNA translation be repressed or enhanced ? What is the half life of various proteins which act as genetic repressors and inducers (if these be protein) ? What is the probability that a single repressor (inducer) molecule will attach to a target gene per unit time ? How tightly do repressors (inducers) bind to target genes, that is, what is the probability of their release per unit time ?

From among these factors, the simplest generalization from binary deterministic nets to nets whose components could be Nary or continuous would seem to be that in which the probability is specified that a single input molecule will attach to a target gene during one time unit (taken equal to the time of transcription of one mRNA), and the probability of gene activity is then computed as a function of the numbers of various input molecules. In this simplest generalization, let us assume that all single input molecules, whatever species, have the same probability, P, of attaching to their target gene in one time unit. Further, let us assume that any attached input molecule is released with probability 1 after one time unit. In addition, it is simplest to assume that all gene product molecules have exactly the same half-life. Finally, it is easiest to ignore initially any distinction between mRNA and protein, and consider the gene to have a single product species which serves as its output molecule and the input molecule to other genes. One is then left with a net composed of devices whose output rates are non-linear functions of their input levels. The purpose of these simplifications is to ascertain whether restricted dynamic behavior in randomly assembled dynamic systems with two inputs per component is limited to nets of binary devices, or is true of a far wider, and more biologically reasonable class of dynamic systems.

Two models have been considered. In the first, the gene is a device whose output is a continuous function of its inputs, fractions of molecules are made and lost, and the system is a deterministic, dynamic net. In the second, the gene realizes a probabilistic Nary function on its integer number of input molecules, and the system is a stochastic dynamic net in which, at each time moment, only an integer number of molecules are made or lost. In both cases, the genome consists of N genes, each of which receives exactly two inputs chosen at random from among the N. Each gene is assigned at random one of 16 possible functions on its two inputs. Once constructed, net structure remains fixed. Every product molecule has an identical probability, P, of attaching to its target gene in one time unit. Each gene is free of input molecules at the end of each time unit, and able to respond to new input conditions. Every product molecule has the same probability, *ploss*, of being destroyed during each time unit.

The 16 possible functions are generalizations of the 16 Boolean functions of two inputs. They are formed as follows : A single input molecule has some probability, P, of finding its target gene during one moment, and a probability of $(1 - P)$ of not finding it. If there are X input molecules, the probability that none will find the gene is $(1 - P)^X$, and it is $(1 - (1 - P)^X)$ that at least one will find it. The 16 functions are built up from these components. Thus, $(1 - P)^X$ is the probability that a gene with one repressor input gene will make a product molecule if there are X input molecules. $(1 - (1 - P)^X)$ corresponds to an inducing input. $(1 - (1 - Pa)^{Xa}) \cdot (1 - (1 - Pb)^{Xb})$ is the generalization of the *and* function for a component with inputs A and B. Values of P and *ploss* have ranged from 0·3 to 0·1, and from 0·1 to 0·033, respectively, in these studies.

Continuous element nets

▶ *Continuous deterministic model* (*CD model*). In the continuous model, fractional product molecules are allowed. In computer simulations, at each iteration, and for each gene, the probability R that a product molecule is produced is computed, then exactly R of a molecule is produced and added to that gene's product pool one moment later. Also, exactly *ploss* of the molecules in each gene's product pool are lost at each iteration. The system is therefore one of continuous, nonlinear, deterministic components, each receiving two randomly assigned inputs, and realizing one of 16 randomly chosen generalized Boolean functions on those inputs.

In computer simulations, a random initial distribution of size of product pools is chosen, and the net is followed through many iterations. Oscillations in R and pool size are followed. In these nets, with N ranging from 4 to 200, the systems quickly settle down to a single steady state level of activity and pool size for each gene. About 100 nets of various sizes have been simulated. In no case so far has any net been demonstrated to have more than a single steady state.

Although one net possesses a single steady state, it may be capable of oscillating about it in more than one way prior to complete damping. Once the steady state values for a net are known, on subsequent runs with the same nets, the 'phase value' of a gene's activity may be defined as 1 if R, the probability of making a product molecule, is above the steady state level for that gene, and 0 if equal or below it. A net of N elements therefore has 2^N phase relations. As the net iterates, it settles into a recurring sequence of phase relations (one mode of oscillation about the steady state), which gradually damps. With a 10 element net, twenty runs from different initial states in the same net generally reveal from one to three distinct cycles of phase relations. Thus these systems can behave in different ways until their oscillations damp to the same steady state. The expected number of modes of behavior as a function of N is not yet known. It may be noted that phase relation cycle is the generalization of state cycle in binary element nets. In these binary element nets, the expected number of state cycles was about the square root of N. About the same result may be found in these continuous, deterministic nets.

Behavior in these nets is highly localized. That is true, of course, in that they seem to have single steady states to which they return after perturbation. More significantly, the number of phase relations which recur on a cyclic mode of oscillation, while the system is not at the steady state, is very few. In nets of 10 elements and 1,024 phase relations, the typical number of phase relations on a cycle is about 10 to 20. In binary element nets, expected cycle length was about

40

the square root of N; the function in continuous nets is not yet known, but may be similar. The results are sufficient to show that the restricted behaviors of randomly constructed nets of binary elements appear to generalize to nets of nonlinear, continuous, deterministic components.

▶ *Integer stochastic nets* (*IS*). The integer stochastic model considered next is the most biologically reasonable of the three models (*B*, *CD* and *IS*). Since the number of repressor molecules per operator locus is low, probably about 10 [1], the activity of a gene is most reasonably viewed as a discrete, probabilistic, not continuous function of its inputs. It seems likely that a good model will be based on the probability that the gene makes an mRNA molecule for each concentration of the various species of its input molecules.

The equations defining an *IS* net are identical to those of the *CD* model, but, in this interpretation, fractional molecules are not allowed. A gene either does or does not produce one whole molecule at each moment, with a probability, *R*, computed from the same probabilistic function used in the continuous interpretation. At each iteration, *R* is computed for each gene from the integer number of molecules in its input pools, then it is randomly decided according to *R* whether a molecule is produced. If so, it is added to the gene's pool one moment later. Also, it is randomly decided according to *ploss* whether each molecule already in the gene's product pool is destroyed. Thus, only integer numbers of molecules are made, or lost, and the system is stochastic.

The steady state values of *R*, found in the *CD* interpretation of the same net, are used to define phase relations in the *IS* interpretation of the net. If its rate of activity, *R*, is above the steady state rate, a gene's 'phase value' is 1; if below or equal, it is 0. As the net iterates, the program compiles a record of the number of times the system is in each of the possible phase relations. Nets with 4 elements and 16 possible phase states occupy, on the average, 8 phase states in 60,000 iterations. The average total number of states occupied by a net of 10 elements and 1,024 states is 224. The behavior of the larger nets is proportionally more localized than the smaller. This is clearer if the localization of behavior among the occupied states is considered. Nets of 4 elements spend half their time (on the average) among about 6 phase states. Nets with 10 elements spend half their time among about 28 phase states; about 3% of the 1,024. It seems likely that larger nets will be proportionately even more restricted; however, the shape of the curve is as yet unknown.

Because the steady state size of product pools is rarely an integer number of molecules, the integer stochastic model can never reach the steady state, and

41

continuing oscillations about it are guaranteed. The rate of activity of most genes oscillates in a narrow range, with only a few genes showing a wide oscillation.

The *I S* nets can behave stably in more than one way. If the same net is run from different initial states, it can be shown that the net occupies different subsets of phase relations. Runs from 20 initial states generally reveal 1 to 3 different modes of behavior. These differences are sometimes stable. A net with several modes may remain in 1 mode for 60,000 iterations. Other nets may move from mode to mode, perhaps 3 times in 60,000 iterations. The average level of activity of a gene can vary. Often, a gene is active in some modes and stably zero on others.

▶ *Discussion*. Very restricted modes of oscillation, discovered in randomly built nets of binary elements, are also found in nets of continuous or probabilistic components when each component receives two inputs. Orderly behavior in randomly constructed dynamic nets is to be expected. To the extent that this is so, these findings must color our attitudes about the difficulty natural selection had in finding 'workable' homeostatic systems. The clear suggestion is that nearly any dynamic biochemical system in which each component is affected by only two or three other components would show restricted, homeostatic behavior. It seems particularly significant, therefore, that most genes subject to transcription control, which have been studied, have about two species of control molecules. Allosteric enzymes also seem to have very few species of control molecules. Quite possibly, the high molecular specificity of macromolecules suffices to explain both precise product formation and, due to the low number of inputs per component, restricted global oscillatory behavior sufficient for homeostasis.

The extent to which very large integer stochastic dynamic nets of the type modeled show localized behavior is not yet known. Nets of 10 components spend half their time on about 3% of the possible phase relations. It will be important to study larger nets. In anticipation of those results, it seems possible that the unsuspected localized behavior of randomly constructed dynamic nets is an important feature of both cellular homeostasis, and perhaps the possibility of evolution itself.

▶ *Cellular Differentiation*. Factors controlling differentiation may be roughly grouped, according to the time required for their reversal. At one extreme is the irreversible loss of genetic material in such organisms as Ascaris, or permanent repression of a portion of the genome as in, perhaps, the repressed X of a female mammal. At the other extreme is the fluctuating activity of a gene which is readily repressed and derepressed, for example *Beta* Galactosidase in *E. Coli*.

S. Kauffman

Relatively stable alteration in the activity of a gene, for example, by its firm repression, its endo-replication, or proliferation of its long-lived mRNA, may be thought of as altering the parameters of the more rapidly fluctuating dynamic system, comprised of easily repressed and derepressed genes, their products, and their interaction. With the parameters fixed at a set of values, it has been suggested, for example, by Kacser [2], that further cellular differentiation may be due to the occurrence of multiple steady states in the behavior of the rapidly changeable dynamic system.

While valuable, the concept of alternate steady state behavior in an open thermodynamic system is both factually and theoretically inadequate to describe the cell; factually, since the levels of various components in a cell maintain oscillations [3]; theoretically, since steady state behavior is a limiting case of the more general stability concept, limit cycle behavior.

It would seem wise to base a theory of cellular dynamic behavior on this more general concept of restricted modes of oscillation in the rates of change of system components. The important feature of the models presented above is their possession of stably different, restricted modes of oscillation. This indicates that, with a given set of parameter values, cells may differ by behaving on different modes of oscillation. Indeed, the common occurrence of multiple modes of oscillation in randomly constructed integer stochastic nets suggests that this characteristic behavior may prove to be among the most common causes of differentiation.

A particular strength of the approach utilized is that quantitative estimates of the number of modes of oscillation may be obtained. Although it may prove factitious, it is of interest that the established number of modes of behavior of binary element nets per net size predicts roughly the number of cell types in an organism per genome size. The expected number of modes of oscillation per net size is not yet known for integer stochastic systems of the type described.

The surprisingly restricted behavior of randomly constructed nets with two inputs per component suggest both further experiment and a technique for modelling cellular dynamic systems. It should be possible to ascertain experimentally the average number of direct inputs per gene. According to this model, the average should be very low. As the average number of inputs, the distribution about that average, and the frequency with which different functions are realized by components become known, the behavior of typical nets built according to those averages can be studied. This may allow insight into the behavior of a complex cellular dynamic system before the entire details of its construction are known.

43

References

1. M. S. Bretscher, *Nature 217* (1967) 509.

2. H. Kacser, in (C. H. Waddington, ed.) *The Strategy of the Genes* (Allen & Unwin: London 1957).

3. B. C. Goodwin, *Temporal Organization in Cells* (Academic Press: London 1963).

Comment by J. Burns

Binary element nets. 1. We suspect that for $K = 3$ the median cycle length rapidly increases for $N > 50$. Thus for $N = 128$ we found only 4 cycles satisfying (RUNIN < 510, CYCLE $\leqslant 256$) out of 25 trials on each of 10 networks.

However we believe also that a limited proportion ($< 50\%$) of the 'genes' *can* have higher connectivity ($K \geqslant 3$) without the results for $K = 2$ breaking down completely.

2. If tautology, contradiction and Booleans which are essentially 'one connected' are excluded, we are left with a 4 : 1 ratio of 'AND' to 'EXCLUSIVE OR' type Booleans. We have confirmed the result of Walker and Ashby [6, p. 36], that is, that 'EXCLUSIVE OR' alone gives immense cycles and that 'AND' alone gives short 'Kauffman type' cycles.

We have also tried introducing an increasing proportion of 'EXCLUSIVE OR' into an 'AND' network and find that approximately 60% of 'EXCLUSIVE OR' can be tolerated before Kauffman type behavior breaks down.

3. In view of the above it seems that two general conditions would have to be satisfied before simple behaviour could appear in a binary net.

(a) That more than 50% of 'genes' have only two inputs.

(b) That more than 50% of control functions are essentially of the 'AND' type. These conditions are implicit in the Kauffman model. Have we evidence that they are satisfied in real genetic systems? Regarding condition (a), even if we assume that most structural genes are controlled directly by a single type of apo-repressor and its corresponding co-repressor, we have to admit also that *many other* gene products will determine, via the faster time scale of intermediary metabolism (Kauffman, p. 43), the level of the small molecule co-repressor. Do we therefore have a single major input and a number of other inputs, quantitatively becoming increasingly less important with metabolic distance from the co-repressor?

Regarding the average nature of the 'control function' the 'EXCLUSIVE OR' would seem unlikely for those genes controlled by a protein and a small molecule

44

since it would imply that the *small molecule alone* could control gene activity! Thus essentially only 'AND's would remain and condition (b) would be satisfied. If, however, the inputs are considered as two proteins 'EXCLUSIVE OR' would be possible. Are there any definitive molecular biological systems where a structural locus is specifically affected by *two* regulator genes?

▶ *Continuous element and stochastic element nets.* The behaviour of the continuous deterministic model is most intriguing, that is, that it has a unique stable steady state.

However, its relation to the original binary model is not simple. The introduction of the continuous generalisation of a Boolean seems OK, although it may be that the functions chosen are too smooth and should have something more like a step in them?

The introduction of gene-product pools with a decay constant amounts to making the input to other genes a weighted average of previous states of the effector genes and may have an important stabilising effect.

The stability properties of a model in which the activity of a gene at $t+1$ is a continuous Boolean function of the activity of its input genes at time t would clarify the question of whether the stability found is due mainly to the continuous Booleans or merely to the introduction of a decaying pool effect.

In the same connection networks with continuous Booleans corresponding only to 'EXCLUSIVE OR' should have the best chance of being oscillators!

The point raised on p. 41 about the integer–stochastic model would seem to be relevant to the deterministic binary model which is also constrained to have an integer solution.

Comment by C. H. Waddington

Is it at all possible to relate these Boolean connecting-functions, however speculatively, with the characteristics we know of in the genetic systems of organisms which really differentiate — namely higher organisms? The most exciting, because the most unexpected, discovery of molecular biology in the last few years has been that the genomes of eukaryotes contain stretches of DNA which are reiterated, identically or nearly-identically, very manyfold, even up to a million times in the most extreme cases. The function of this reiterated material is still almost entirely unknown, except in the one case of the ribosomal cistrons, where the reiteration can plausibly be connected with the necessity at certain stages of development (for example, in oocytes) to synthesize a very large

Comment by C.H.Waddington

amount of ribosomal RNA in a very short time. Speculation is still at liberty to ask what questions it pleases about most of the rest of the reiterated DNA. Now there are some indications that the RNAs transcribed from this reiterated DNA never get out of the nucleus (they may be 'Henry Harris RNAs') ; and if so, one might suspect that they are concerned with gene control rather than with the production of protein-effectors. Now one might argue that the agents we know of in the nuclei of higher organisms, and can call on to explain the control of the activity-state of various genes, appear to lack sufficient specificity to do the job by anything like a one-to-one reaction ; but the situation would be saved if the control–acceptor sites associated with each structural gene were sufficiently redundant to compensate for the high noise level in the controlling inputs. The stretches of DNA reiterated from a few hundred to a few hundred thousand times may be like the letters of the alphabet, which have to be re-iterated many times if they are to spell out a set of complicated paragraphs of 'control-accepting instructions'. And the less specificity ('information') is carried in the incoming controlling agents, the more complex the receiving-instructions must be. Might it be, perhaps, that the theory we really want is one which is 'anarchic' not only in that the connections between effectors (genes) is made at random, but also in that the connections are not simple clear-cut Booleans, but are 'noisy Booleans' – i.e., correspond more nearly to complex control-accepting elements connected to the genes ?

The synthetic problem and the genotype-phenotype relation in cellular metabolism

J. Burns
University of Edinburgh

Introduction. The basic problem to be discussed in this paper is how our knowledge of cellular components can be used to gain insight into quantitative aspects of theoretical synthetic systems constructed from them. It is the quantitative phenotype, arising from the genotypic prescription and the environment, which is of critical importance for the cell's survival and which therefore features in population genetic theory. A study of this 'synthetic problem' would thus, by providing genotype–phenotype mappings for simple synthetic systems, help to connect two major areas of biological theory : the biochemical and the population genetic.

Such a study would be seriously complicated, or even rendered not feasible, if the long-term behaviour of synthetic systems were commonly oscillatory or such that many different steady states could arise from a given genotype and environment. However Griffith [1] has shown recently that simple repression and induction loop models possessing one enzyme have, under most circumstances, just one stable steady state and no limit cycle behaviour. Our own computer studies show that this result remains true for a single pathway containing several enzymes under co-ordinate control, and also for more complicated networks of intermediary metabolism, but without repression or induction loops. On the other hand intermediary metabolic oscillations can certainly occur [2].

It would seem justified to make a working assumption that commonly the steady state concept will be relevant but that an occasional 'oscillator' will be encountered.

For many purposes it is possible to avoid the 'gene to protein function' problem and to start from the proposition that 'a genotype with n loci is equivalent to a functional description of (each of) a set of n proteins'; it is in this sense that 'genotype' will be used in what follows. A mutation at a locus is then represented by an altered description for the corresponding protein, and alternative alleles at a locus determine which of alternative descriptions is to be used. Genotypes possessing loops to control both enzyme quantity and activity will correspond to suitably described allosteric effects being allowed for in the set of proteins.

47

Cellular metabolism

This approach allows a variety of theoretical questions to be considered. Thus one could ask in cell physiology, for a given genotype, what the effect of a mutation which abolished a repression loop would be, or in population genetics what proportion of the variability would be additive in a specified population of genotypes containing known alternative alleles at each of n loci.

As molecular biology refines its description of cellular machinery, this ability to ask meaningful theoretical questions will increase. The ability to answer them will depend on the degree to which the 'synthetic problem' can be solved. Thus a clear challenge is presented to 'theoretical biology'. In the remainder of this paper an approach to the synthetic problem will be discussed and an application in cell physiology outlined.

▶ *The synthetic problem.* The recent successes of molecular biology in isolating and determining the function of many cellular components has made more acute the dilemma that 'theoretical biology' has no simple method for dealing with the synthetic problem'; that is, the problem of the quantitative prediction of systems composed of large numbers of different but, in principle, well described components.

The successful system theory that exists appears instead to be of a qualitative or even 'logical' nature and perhaps owes its success to the fact that it has avoided the quantitative morass and dealt with well-defined logical questions such as 'coding' or the positioning of enzymes in a metabolic pathway. At the level of cellular metabolism it is usually possible to predict, given a relational map of metabolic components, the *direction* of response to some change but, even here, if several components are altered prediction becomes impossible.

Answers to the 'synthetic problem' are divergent, varying from Watson [3] who is optimistic that we can discover quantitative aspects of a system whose components are given, to Elsasser [4] who appears to believe that there are basic and important indeterminants.

▶ *Computational approach.* The approach adopted here is not to aim for a precise mathematical model of any aspect of metabolism, but rather to use models embodying rough quantitative descriptions of the relevant cell components and producing quantitative behaviour which is roughly similar to the real system. That this is a valid approach can be seen from the fact that in a biological system quantitative behaviour depends firstly on the structure of the system and only secondarily on the precise performance of any component [5, 6].

Each protein, then, owns a 'rate expression' defining its action and how the rate of this action is to be calculated, at any instant, given the concentrations of small molecules having strong interactions with the protein.

The instantaneous rate of change of any metabolite (small molecule) will be the algebraic sum of all rate expressions affecting it and thus a system of first order ordinary differential equations can be set up to represent the temporal behaviour of the system. These equations are non-linear and not amenable to analysis. It should be noted that no attempt has been made to include the spatial organization of the cell.

Given that the system possesses a stable steady state the problem becomes that of solving numerically a bank of non-linear equations to determine the metabolic values corresponding to the stationary state. This is done by first integrating the equations of motion of the system to get into the neighbourhood of the stationary state, and then converging to an exact steady state using a method akin to Newton–Raphson. This has the advantage that 'oscillators' can be detected and their frequency of occurrence noted. The method is described more fully elsewhere [7].

Thus it is now possible, subject to the limitations discussed, to move from a 'genotype' to its corresponding 'phenotype'. An example where this can be applied will now be outlined.

▶ *Optimum growth rate problem.* A general aspect of cell physiology is studied here at a theoretical level. If one considers an exponentially growing metabolic system contained at time t in a volume V and having steady enzyme concentrations $E_1 \ldots E_N$ then one can show that

$$\text{exponential growth rate} = \frac{1}{V}\frac{dV}{dt} = \frac{F(E_1 \ldots E_N, X_1, X_2 \ldots, P_1 \ldots)}{(E_1 + E_2 \ldots)}$$

where F is a function giving the net rate of protein production from the system per unit volume in terms of $E_1, E_2, \ldots$, the enzyme concentrations, $X_1, X_2, \ldots$, the concentration of nutrients in the environment, and $P_1, \ldots$, the kinetic constants of the enzymes.

This growth rate will be a maximum when $\dfrac{E_i}{F} \cdot \dfrac{dF}{dE_i} = \dfrac{E_i}{(E_1 + E_2 \ldots)}$ for all i,

that is, when a particular 'allocation' is made of the protein production F. This ideal allocation will depend on the environmental parameters $X_1, X_2 \ldots$. Thus in different environments the organism cannot maintain its optimum growth rate with a fixed allocation scheme. A suitable scheme of inductions and

49

repressions controlling the 'relative' rate of production of the different proteins can adjust the allocation as the environment changes.

Using the computer model of the metabolic system one can observe for each environment the growth rate for a fixed allocation, that when control loops are added, and finally that for the computer-calculated optimal allocation. This enables one to study for a range of metabolic networks the effectiveness of various patterns of control in achieving a stated objective.

▶ *Discussion.* No mention has been made so far of the computer time needed for these computations, although this has an important bearing on their feasibility. Approximately 10 seconds is required to locate the steady state of a typical 15 enzyme system using an IBM 7094 computer, but this figure can vary considerably with the parameters of the enzyme system.

This means that physiological studies, which might involve a number of parameters being moved over a range to discover response curves, are computationally feasible, taking only minutes or tens of minutes.

Synthetic metabolic systems, of the type described earlier, certainly possess, at the physiological level, such phenomena as epistasis, dominance, and pleiotropy, and these arise as natural properties from the molecular assumptions. It would seem interesting to consider these synthetic systems as objects for study using the methods of population genetics.

However, computational difficulties arise with population genetic studies. A Monte-Carlo simulation of a 10 locus problem with an offspring population of 20 animals would take about 3 minutes / generation compared with 1 / 10 second for a symmetric non-epistatic model on the same machine. The situation can be improved by holding a pre-computed table of genotypic values, but this method becomes impracticable at about 6 loci. Another possibility is that a 10 locus problem, having $3^{10} = 77,049$ physiologically distinct genotypes, could be handled by computing a few hundred of its genotypic values and using them in a table to predict others as required; indications are that this is not a simple problem.

Perhaps the best approach to population genetic studies will be to consider the nature of the genotype-phenotype relation at the physiological level in the hope that some relatively simple rules for compounding genetic effects operating through a metabolic network can be discovered.

This work was carried out while the author was a member of the Epigenetics Research Group, Edinburgh.

J. Burns

References

1. J. S. Griffith, *J. Theoret. Biol. 20* (1968) 202.
2. B. Chance *et al.*, *P.N.A.S. 51* (1964) 1244–51.
3. J. D. Watson, *Molecular Biology of the Gene* p. 100 (Benjamin: New York 1965).
4. W. Elsasser, *The Physical Foundation of Biology* (Pergamon Press: Oxford 1958).
5. H. Kacser and J. A. Burns, *Quant. Biol. of Metabolism.* Third Symposium (Springer: New York 1968).
6. J. A. Burns, *Quant. Biol. of Metabolism.* Third Symposium (Springer: New York 1968).
7. J. A. Burns, *FEBS Letters 2*, Suppl. S30.

Evolution of genetic conformation

Alex Fraser
University of Cincinnati

Evolution in a non-sexual organism depends on a sequence of mutational events occurring in a single lineage of reproduction such that the accumulated effects of the mutations result in an increase of the probability of survival of the lineage. A strong case can be made for this statement being invalid since known rates of mutation and known amounts of chemical evolution are not reconcilable. The probability that such amounts of chemical evolution would occur, even over the long ranges of evolutionary time periods, is so infinitesimally small that the above statement of evolution cannot be accepted; there is some factor missing.

The problem can be phrased in terms of evolutionary changes of a single gene. This gene, a single length of DNA, will be varied by mutational changes of its elements, the nucleotides. The evolutionary change of this molecule will depend on the sequence of accumulation of such mutants resulting in an increase of the probability of survival of the lineage of reproduction of the gene. This rephrasing in molecular terms allows a more precise rephrasing in quantitative terms.

A vector of j elements, with each element having one of four possible values, is copied. The copy is then taken as the basis for the production of a second copy. This is repeated n times. If miscopying can occur such that each element of the vector has a probability, μ, of being replaced by any one of the three alternative values in the copy, then what is the pattern of requirements for a particular vector to be transformed by repeated miscopying into another vector where k of the elements differ between the initial and the final vector specifically? If we consider just one lineage of replication of the vector, then for j, k and μ having values which are known to occur, then n approaches infinity. If we consider m separate lineages of replication of the vector, then for m too large to be accommodated on earth, n is still too long to be considered as within evolutionary time periods.

Since evolution has occurred it follows that the above statements are wrong. We can see where the statement is wrong by considering evolution in a sexual organism. Evolution in a sexual organism depends on mutational events occurring in a set of lineages of reproduction such that the accumulated effects of the mutations result in an increase of the probability of survival of the set of lineages. The critical difference of sexual from non-sexual reproduction, is that

52

the set of lineages are connected by the combination of genes from two lineages allowing recombination to occur, initiating a new lineage descendent from two of the preceding lineages.

The term 'sexual' is used loosely to include all forms of genetic combination and recombination both obligate — occurring in the life cycle of each individual — and stochastic — occurring with a probability of less than unity in the life cycle of each individual. A very small amount of sexual reproduction appears to be sufficient to transform the genetic structure from the independent parallelism of non-sexual reproduction, to the reticulate interconnection of sexual reproduction.

The critical feature is not so much the combination of genes as their recombination. Crossing over can result in the production of a gene containing two separately located mutations, each having occurred in a separate lineage of reproduction. If the probability of such a recombinational process is sufficiently high then the probability of the evolution of known changes of conformation of genes becomes sufficiently large to be acceptable.

There is a need for studies of the relation of mutation rates, recombination rates, and population size to the probability of occurrence of specified genetic conformations. This problem has proved to be mathematically intransigent and recourse has been made to computer models. These suffer from the defect of restrictions of the size and duration of the models that can be examined. Economic restrictions set limits to the length of the genetic vectors, to the number of lineages of reproduction, and to the number of generations of reproduction. It has been necessary, therefore, to examine situations in which information can be expected to be produced over fairly short time periods. Such a situation is one of obligate sexual reproduction in a diploid organism. The great majority of computer models of evolution have been based on diploid obligate sexuality.

A model we have used involves a genetic vector of j elements, with j having values between 2 and 30. A rate of recombination between adjacent elements, r, having values between 2^{-1} and 2^{-10}, and a population size of 2^5 to 2^{10} lineages of reproduction. Each element of the vector has two states : 0 or 1. The effect of mutation was included by specifying that all possible genetic conformations had an equal probability of occurring in the initial state of the population. This is the state that a population would be in if mutation had the same value for all possible changes of state, and if there were no differences of survival value between genotypes. This could be expected to occur if the genetic locus had been suppressed from having any phenotypic effect for a long period. Removal of the suppression will result in the gene having an effect on the phenotype, and selection will be operative. The model was based on a particular mode of selection that has been variously called normalizing, optimizing, or stabilizing selection. The value of the genotype is determined by the simple sum of the elements of the two genes, and selection involves individuals with intermediate values of this sum having a greater probability of transmission of their genes.

The results of computer runs of such models depend markedly on the size of the population, the number of elements in the gene, and the rate of recombination between elements. If the rate of recombination is high (2^{-1} to 2^{-4}) and the size of the population is large, then selection is ineffective. Under conditions of a significant genetic drift, that is, where a small population size introduces a strong chance-determined variation of genotype frequencies, selection is effective, resulting in the fixation of the population on one of the balanced combinations. If $j = 6$ there are twenty such combinations, for example, 000111, 111000, 101010, 010101, and all of these are equivalent in terms of the probability of survival of the population. Selection in this situation results in the population becoming fixed for 1 out of 20 equivalent genetic conformations, each occurring with a probability of $1 / 2^{j}$ in the initial population. The dependence of evolution of a balanced genetic conformation on the occurrence of a genetic drift situation increases as the number of elements increases. The need for a strong genetic drift comes from the very large number of combinations that are genetically balanced, and have the same probability of genetic survival. A chance factor is needed to perturb the genetic situation such that one of these alternatives becomes pre-eminent. This is extremely evident from computer runs.

The situation where the rate of recombination is low (2^{-4} to 2^{-5}) is very different from that in which the rate of recombination is high (2^{-1} to 2^{-4}). Here we find that there is consistent increase of frequency of two particular genetic conformations : 0101 . . . 01 and 101010 . . . 10. This increase occurs under conditions where selection is otherwise ineffective, that is, in large populations for genes with a large number of elements. Here we have two genetic conformations becoming a major component of the genetic constititution of a population in which initially their frequency, $1 / 2^{j}$, was so low that they could be considered as effectively non-existent, for example, in computer runs made for $j = 30$ where the size of the population was small. In actual computer runs with $j = 30$, the two fully balanced conformations were not present in the initial populations, appearing only after a number of generations of selection. Selection resulted in the construction of the two fully balanced conformations by recombination of parts of unbalanced conformations. The reason for the increase in frequency of the two fully balanced conformations is that these have an advantage in the characteristics of the range of progeny produced from individuals having such conformations. The range is smallest for progeny from fully balanced conformations.

The examination of these models has shown that selection can result in extensive restructuring of the genetic conformation over very small periods of time (a hundred generations) and in very small populations (a few hundred individuals). Clearly, there is much greater scope for events of even lower probability being evolved in the much longer periods of evolutionary time (thousands of generations) over the much larger range of populations (thousands of individuals).

A criticism of this model is that it is based on the initial population being at mutational equilibrium in the absence of selection; a state that can be considered to be fairly rare, if it ever occurs at all. We know that natural populations show a very high degree of polymorphism, for example, the haemoglobin variants, where selection is operative. Explanation of this in terms of heterozygous advantage seems unlikely, and it is probable that the final explanation will involve frequency dependent advantages, but until this problem is resolved it is more realistic to base models on initial states of mutational equilibrium; less assumptions are involved.

More work is needed, particularly in the area of stochastic sexuality, but it is unlikely that the general conclusion will be changed; namely, that the missing factor in evolution in non-sexual organisms is genetic combination and recombination. We need to reconsider any models of evolution that are premised on an initial phase of non-sexual reproduction in terms of the genetic mechanism being inherently combinational and recombinational, that is, these phenomena are not evolved, but intrinsic in the initial chemistry of life.

The requirement for a biologically real rate of combination and recombination of the original life molecules sets specifications for the structure of such molecules. If the original molecule were a single stranded nucleic acid, then its structure would have to be such as to have a high rate of double strand formation and separation. The kinetics of cross-strand pairing would have to have lower thresholds than in present nucleic acids. Further, the stability of the nucleic acid backbone would have to be lower than in present nucleic acids such that recombination between strands would occur with a biologically real probability. The conclusion is that the initial nucleic acid molecules would need to be strongly divergent from present types, which can be considered to be highly evolved. A new area of theoretical biology that would seem worthwhile is the invention and examination of 'primitive' nucleic acids.

References

References to work on computer models of genetic systems are given in Fraser and Burnell, *Computer Genetics* (McGraw-Hill: New York 1970).

An epigenetic system

Alex Fraser
University of Cincinnati

Geneticists have had, essentially, two main fields of preoccupation : the problem of the structure and replication of the gene, and the problem of the structure and homeostasis of the genotype. It is now apparent that the first problem is essentially solved — the gene is a molecule of DNA, whose replication is based on the complementarity of the two strands of the double helix, and mediated by a system of DNA polymerases. The problem is one of detailed biochemical interest rather than general genetic interest. Attention is returning to the problem of genetical homeostasis, catalysed by the work on regulator–operon systems. It is now apparent that the total genotype is separable into subsets that are closely related, not so much in their primary actions as in the occurrence of interactions which integrate the actions towards a single component of the phenotype. Such a subset can be called a genetic system. The ribosomal cistrons and the transfer-RNA cistrons are each central parts of a genetic system, whose function, translation of messenger-RNA, is critical to the functioning of the living mechanism.

These two fields of preoccupation relate fairly closely to the older concepts of preformation and epigenesis. The preformed basis of an individual has its identity in the constant informational content of the genes (the homunculus has a nucleic acid morphology). The epigenetic translation of the genetic information involves complex sets of genes acting in a variable milieu of genetic and environmental effects, such that constant progressions and fixed end points eventuate.

Kauffman (p. 18) has investigated the properties of models of genetic systems, finding that if the number of interactions per gene is restricted to a small number, then large genetic systems undergo cyclic patterns of change that are surprisingly short. It appears that a large genetic system may have a relatively simple pattern of behavior. Kauffman's models are several stages of extension beyond present descriptions of actual genetic systems. Burns (p. 47) has begun the simulation of genetic models that are more easily identified with known systems.

The literature is crowded with descriptions of the genetic systems controlling biosynthetic pathways, but very few of these are obviously canalized, and Waddington has emphasized that a profitable area of study is that of genetic systems controlling strongly canalized characters, since canalization involves complex

57

patterns of interactions between genes. A considerable effort has been directed at the genetic analysis of characters which exhibit a lack of variability based on a variable genetic system, that is, of characters that are strongly canalized.

The first stage of analysis of such canalized characters has been based on models of a system of genes and environmental effects that interact to determine the potential value for the character (the 'make' in Rendel's terminology[1]) and another system of genes that act to determine the relation between the potential value and the final phenotype. We need to consider as an evolutionary starting point the absence of the second system of genes, that is, a simple identity between the potential and the final measure (figure 1). Selection acting to stabilize the phenotype at some norm, can be expected to act directly on the genetic determination of the measurement, restructuring this to a balanced polygenic or homozygous format (see Fraser, p. 54), if the variation of the character is predominantly genetic with little involvement of the environment. If the variation involves an environmental component, such stabilizing selection can achieve a reduction of variability by the structuring of a separate genetic system which suppresses the expression of the potential measurement. It can result in the formation of a system of canalizing genes which act to reduce or enhance the effectiveness of the potential measurement, dependent on the direction of the deviation of this potential value from the norm. This latter system is the only method of establishing a constant norm where environmental factors are predominant.

Such a two-part genetic system, a set of genes determining the potential measure, and another set of genes modifying the range of expression of the first set, will not be enough to suppress fully the variability of the potential measure. The second set of genes – the canalizing system – will need to have two parts. The first part will act to suppress the potential measurement increasingly as the potential measure increases. The second part will act to desuppress this suppression increasingly as the potential measure increases. Such a three-part genetic system can be balanced to result in a function relating potential measure to final phenotype, which will not result in increase of the final measure over a considerable range of potential measurements. Fraser (p. 56) in a computer simulation of such models found that it is difficult to model systems of canalizaation except on a three-part model. If the system determining the potential measurement has a repressive effect on the suppressor scheme, then only two components are needed – the need is for a closed loop of feed back repressions.

The invention and examination of the properties of models of biological

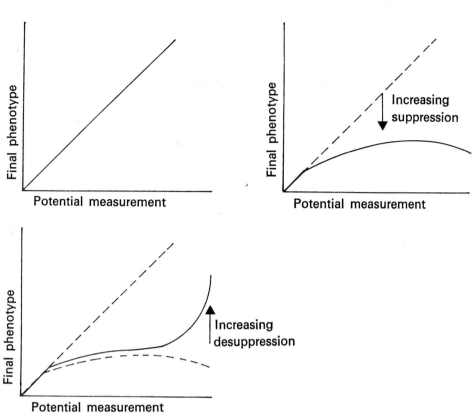

FIGURE 1

systems is theoretical biology, but this needs to be referred against real systems. The genetic analysis of a canalized character, a chreod in Waddington's terminology, has been a major occupation of our group for some years. The character, number of scutellar bristles in Drosophila, is normally invariant at four; only a small percentage of individuals have more or less than this norm. This norm of four is characteristic of the Drosophilidae, and it can be considered to be a feature of the archetype of this group. Selection experiments of various degrees of sophistication have produced strains with numbers of scutellar bristles deviating very widely from the norm. A very noticeable feature of such selection lines is their variability — such lines are not stabilized against the effects of environmental factors. It is apparent that the phenotypic constancy of number of scutellars is based on a suppression of a highly variable underlying genetic system.

Two major loci have been involved in the genetic analysis of scutellar number : scute (*sc*) and extraverticals (*xvt*). Mutants at the scute locus reduce the

An epigenetic system

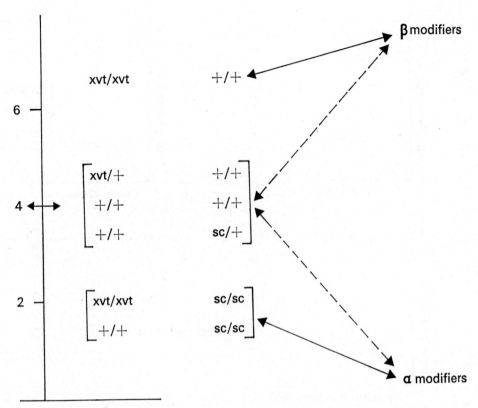

FIGURE 2
The reciprocity of degree of expression of the α and β sets of modifiers in relation to the expression of the *xvt* and *sc* loci. (Continuous lines, enhancement ; dashed lines, suppression.)

number of scutellars, and mutants at the extravert locus increase the number of scutellars. The two most used mutant alleles used in our studies have been sc^1 and *xvt*. The former is a sex-linked recessive, and the latter is an autosomal recessive with a variable penetrance in its effect on scutellars. Scute 1 is a suppressor of *xvt*, that is, $sc^1/sc^1//xvt/xvt$ does not differ from $sc^1/sc^1//xvt^+/xvt^+$ in having a reduced number of scutellars. This feature of suppression of *xvt* by *sc* has been used to construct experiments allowing the identification of different sets of genes affecting the number of scutellar bristles, that is, allowing the dissection of the genetic system for number of scutellar bristles.

Two systems of modifiers of numbers of scutellar bristles have been identified whose degree of expression is reciprocally related to the status of the *xvt* and *sc* loci. The α set is maximally expressed in the presence of sc^1 expression, and suppressed in the presence of *xvt* expression. Conversely, the β set is maximally

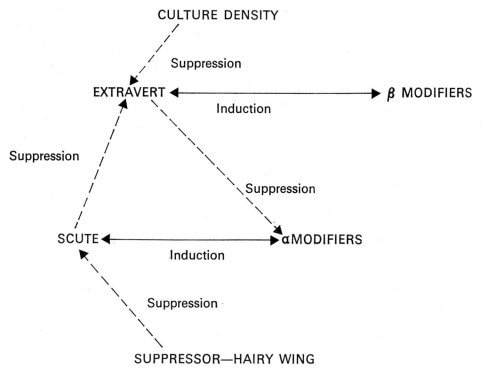

FIGURE 3
An epigenetic network for determination of the number of scutellar bristles.

expressed in the presence of *xvt* expression and suppressed in the presence of *sc*[1] expression. (*See* figure 2.)

These results indicate that an incompletely closed loop of induction–suppression is involved in the genetic system comprised of the *xvt* and *sc* loci, and the α and β sets of modifiers. The *sc*[1] allele suppresses *xvt* and the β modifiers, and the *xvt* allele suppresses the β modifiers.

A mechanism of canalization needs to involve elements that are sensitive to environmental variation, and can translate this sensitivity into actions that are part of the epigenetic network. An interesting development of our studies has been the discovery that the expression of *xvt* is sensitive to culture density. High culture density suppresses the expression of *xvt*. The sensitivity of *xvt* expression to an important element of the environment, and the effect of *xvt* expression on the α set of modifiers indicates that the *xvt* locus is a point in the epigenetic network where a particular environmental effect is introduced into the system.

61

An epigenetic system

A number of other loci are clearly part of this network : tufted, scutoid, poly-chaetoid, and so on, but none of our studies is sufficiently complete to define the positions of these genes in the network. Lee (personal communication) has made some studies of the gene, suppressor—Hairy wing (su.Hw), which is a suppressor of scute. He has preliminary results indicating that the suppression of the expression of scute by su.Hw, does not include the suppression of the suppression of *xvt* by *sc*, suggesting that the scute locus is complex with one part acting on the number of scutellars, and the other part acting to suppress the *xvt* gene. (*See* figure 3.)

Studies like these will provide the factual basis for development of models whose examination over evolutionary time periods should give an understanding of the genetic basis for archetypes and the constraints that they impose on evolution.

Refereneces

1. J. M. Rendel, *Canalisation and Gene Control* (Loges : London 1967).

On the irrelevance of genes

R. C. Lewontin
University of Chicago

The two greatest triumphs of the Morgan school of genetics were, first, the demonstration that the chromosomes were the physical counterparts of the hereditary factors of Mendel and, second, that the chromosomes were not *really* the physical counterpart of the hereditary factors of Mendel. Clearly chromosomes were the carriers of heredity and clearly they assorted themselves at meiosis exactly as Mendel's Laws required they should. Yet further work fragmented that unit and showed that the fundamental quantum of genetics was something smaller. Those smaller units did not obey Mendel's Law of Independent Assortment, nor would they, more simply, stay together so that the chromosome as a whole could fulfil its role as Mendel's quantum, obeying the laws of segregation and independent assortment. Rather, it became apparent that the genome was hierarchically organized with quanta of heredity, correlated in their statistical behavior at meiosis, organized into larger units, the linkage groups, which behaved relative to each other in a simple Mendelian fashion. The hierarchy was further complicated by the more recent discovery that even those quanta, Morgan's genes, were themselves subdivisible by the same process of recombination that seemed to define them for Morgan. Thus, our present picture is of nucleotide bases, three of which make a codon; of codons, 200 or so of which make a gene; of genes, 2,000 or so of which make a chromosome; and of chromosomes, a dozen or so of which make a genome. And of all these units of hierarchy, it is only the chromosomes that obey Mendel's Law of Independent Assortment, and only the nucleotide base that is indivisible. The codons and the genes lie in between, being neither unitary nor independent in their behavior at meiosis.

All of this knowledge of the real structure of the genome has made remarkably little impression on population genetics, or at least *theoretical* population genetics. Population geneticists are often accused of having failed to incorporate the findings of modern molecular genetics. But the situation is far worse than that. They have not even incorporated the findings of Morgan. Nearly the entire corpus of literature in theoretical population genetics is written from the standpoint of single Mendelian genes or else genes that all obeyed the law of independent segregation. With the exception of an occasional consideration by Fisher

63

and Wright in a special context, the fact that the genes were on chromosomes was virtually ignored for a long time. The standard textbook in the field has gone through two editions with only a few pages devoted to the dynamics of two genes on the same chromosome.

Since 1956 and especially in the last ten years, there has been an exponentially increasing attention paid to the problems posed for theoretical population genetics by the linkage of genes. This development, which has progressed from the consideration of a pair of genes to models with dozens of loci segregating simultaneously, has had a surprising outcome. As I will show, genes have disappeared entirely from the theory leaving only the chromosomes as units to be considered!

▶ *The problem of a sufficient dimensionality.* The genetical evolution of a population in time can be represented as the movement of 'particle' in hyperspace, the location of the 'particle' giving the genetical composition of the population. At every point in the space there is a vector giving the direction and rate of motion that the particle would have at that point. The history of the population and its future evolution can then be seen as the trajectory of the 'particle' under the influence of the forces represented by the vector field. At some points in the vector field, singularities, the magnitude of the vector is zero, so that if the population lies at such a point it will not evolve. Such points are the equilibria of the system. These may be stable, unstable, or metastable, depending upon the configuration of the vector field in the neighborhood of the singularity. The most important and biologically relevant case is the stable equilibrium where all vectors in the neighborhood of the point are pointing toward it.

Now imagine that we attempt to construct the vector field for some population in a space determined by certain axes, representing different descriptive parameters of the population. With such a vector field we now construct the trajectories that would be traced out by populations starting from different points. We can then ask the question, 'Do any of these trajectories cross each other?' If any do, then at the point of crossing there is an ambiguity about the evolution of the population. In fact, there is more than one vector at that point and we do not have a proper vector field. Another way to put it is that there is information about the motion of the particles that is not contained in the position of the particle. We will say that in such a case the axes determining the hyperspace are not *a sufficient set* of descriptive quantities in which to follow the evolution of the population. We must then search for added variables which will exhaust the degrees of freedom, and add these as dimensions of the hyperspace.

64

R. C. Lewontin

Obviously there is an infinite set of sufficient descriptions of such a system and, in fact, a set of linear combinations of a sufficient set of variables will be sufficient. We will say that a dynamic system has a dimensionality S when S is the smallest number of axes forming a sufficient set for that system. In these terms, the problem of present day theoretical population genetics is 'What is the dimensionality, S, of an evolving population?'

▶ *Correlations between genes.* For concreteness sake, imagine an organism with m loci (cistrons) each having a alleles segregating in a population. At each locus there are then $a - 1$ independent allelic frequencies, since at each locus the sum of the frequencies of alleles must be unity. The classic view of population genetics is that for such a population $S = m(a - 1)$. That is, the evolution of the population can be completely and unambiguously described by the changes in the allelic frequencies at each locus and, moreover, if the forces of natural selection, mutation, migration, and so on are known, then the future composition of the population can be predicted, given only information about the separate gene loci. Under this view, the joint distribution of different loci is given as the product of the frequencies at the different loci so that the frequency of any particular chromosomal combination, say $a_1 b_3 c_2 d_1 e_4 \ldots$ is given by the product of the separate frequencies of alleles a_1, b_3, c_2, d_1, e_4, and so on.

This view was first attacked at a general level by Lewontin and Kojima [1] and Bodmer and Parsons [2] for the case of two loci using equations that are essentially identical with those first derived by Kimura [3] for the special case of mimicry polymorphisms. What these authors showed was that the frequency of chromosomal combination of genes was not always given by the products of gene frequencies in evolving populations, and that a sufficient description and prediction of evolution of two loci involved specifying the frequency of each gametic type. This is equivalent to saying that the dimensionality of the evolutionary space is $S = a^m - 1$ which is, of course, immensely larger than $m(a - 1)$ of the classical picture. The added dimensions are those accounted for by the correlation between loci. Two obvious alternative parametrizations of the dimensionality are (a) the frequencies, g_i, of all the a^m gametic types (less 1 since $\Sigma g_i = i$) or (b) the allelic frequencies at each locus, measures of 2 locus correlations, 3 locus correlations, and so on, up to m-locus correlations, by direct analogy with main effects, first order, second order, and so on, interactions in the analysis of variance. For the case of two loci with two alleles, this second alternative is rather convenient, in that the population can be described in terms of three variables, p_A, p_B and D, the allelic frequencies at the two loci and a

65

measure of so-called 'linkage disequilibrium'. D is related to the correlation between loci by the relation $D = \rho \sqrt{[p_A(1 - p_A)p_B(1 - p_B)]}$.

Lewontin and Kojima [1] and Bodmer and Parsons [2] showed that under certain kinds of natural selection two different stable equilibria were possible : equilibria in which $D = 0$ (when there is loose linkage between the genes), and equilibria in which $D \neq 0$ when there is tight linkage. In cases of simple symmetrical fitnesses the equilibrium *allelic* frequencies are $p_A = p_B = 0.5$ for *both* of these equilibria. That is, p_A and p_B are not a sufficient set of variables since two quite different genetic equilibria are reached with some value of p_A and p_B. These results which were later extended by Felsenstein [4] and most recently by Karlin and Feldman [5] and Bodmer and Felsenstein [6] who showed that there would be a correlation between loci at equilibrium, provided that loci were not additive in determining fitness and that the recombination between the loci was less than an amount determined by the intensity of natural selection. The magnitude of the correlation, however, was fairly small except for very tightly linked genes or for genes under very strong selection. For example, for two loci both showing 10% heterosis, with all homozygotes equally fit, and with total fitness determined multiplicatively between loci, the recombination between genes would have to be less than $r = 0.0025$ (0.25 centimorgans) for there to be any equilibrium correlation between loci. For a recombination of, say, 5%, in this multiplicative model, the homozygotes could be only about 50% the fitness of heterozygotes at each locus if there was to be any correlation. Moreover, Kimura [7] showed that, if selection is weak and linkage very loose, Fisher's Fundamental Theorem holds to a very close approximation. Thus it appeared that, except for very tight linkage or very strong selection, the genes could be looked at separately and the dimensionality of the evolving system was low and manageable.

This view, based on 2 locus models, turns out to be quite wrong when more genes are considered. The first hint that this might be the case was in a series of investigations of five locus computer models by Lewontin [8, 9] and Lewontin and Hull [10]. These investigations showed that genes far apart on the linkage map could be held in correlation with each other by genes segregating between them. Thus, if the critical value for recombination between adjacent genes is 5%, loci 1 and 5 which are 20 units apart on the genetic map will be held together in a correlated distribution by loci 2, 3, and 4 between them. Two other findings were also suggestive. First, the critical value for recombination between adjacent loci in the 5 locus cases was slightly larger (5%–10%) than for two adjacent loci with the same selection intensity in a two locus case. Second, the intensity of the correlation between two adjacent loci embedded in a linkage group of 5 loci is *much larger* than when they are only two genes segregating. For example, in a multiplicative model in which every locus homozygous cuts the fitness of the genotype in half, the correlations between two adjacent loci in the 2- and 5- locus cases are as follows :

recombination between adjacent loci	correlation at equilibrium between adjacent loci	
	2 locus	5 locus
0·01	0·9165	0·9674
0·02	0·8246	0·9294
0·04	0·6000	0·8226
0·06	0·2000	0·5810
0·063	0	0·4858

Clearly the difference is very great for the looser linkages where the correlation in the 2 locus case fall off very quickly as compared with the 5 locus case.

▶ *Realistic models of the genome.* The previous results are suggestive of what will happen when we make a more realistic model of the genome. By more realistic we mean increasing the number of cistrons to the thousands or tens of thousands. Such an increase results in two trends with opposite effects on the correlation between genes. An increase in the number of genes in the model means a reduction of the linkage distance between adjacent loci so that a chromosome with 1,000 genes but a total map length of only 50 centimorgans will have linkage distances between adjacent genes of only 0·005. This will cause an increase in the degree of correlation along the chromosome. On the other hand, the effect of a gene substitution at each locus separately must be very small when there are so many genes segregating for fitness traits. Then the deviation from additivity between loci will become vanishingly small as the same total fitness differential is divided up among more and more loci. This has the effect of weakening the correlation between loci in the chromosome. We must ask, then, what happens as we increase the number of loci more and more, simultaneously weakening the selective effect per locus and strengthening the linkage between loci? The change from 2 loci to 5 loci suggests strongly that the tightening of linkage will have a stronger effect than the weakening of selection interaction, with the result that in a realistic model with many loci, the whole chromosome may be a highly correlated structure and the dimensionality of the evolutionary space will be of the order of a^m.

Ian Franklin and I have recently studied these more realistic models. We have studied both the multiplicative heterotic model, in which it is assumed that all loci are identical, heterotic, and multiplicative in fitness effects between loci, and a truncation model along the lines suggested by King [11] and Sved, Reed and Bodmer [12]. In the latter case we again assume all loci identical and heterotic, but a constant proportion of the population survives with the most heterozygous genotype having the highest probability of being included in the surviving fraction of the population. We have concentrated on models with 36 loci but have compared these with 2 locus, 5 locus, 18 locus and 360 locus cases. For 36 locus models all the features of 5 locus models are confirmed and amplified.

If a finite population begins its evolution with each of the two alleles at each locus segregating and the alleles at different loci combined at random with respect to each other, then every chromosome in a population of moderate size will be different. As the population evolves under natural selection, all the loci maintain an allele frequency of close to 0·5, when selection is symmetrical with respect to homozygotes, except for an occasional locus that may be fixed by chance. However, the number of different *chromosomes* is reduced during the course of selection until only two or three chromosomal types will make up the bulk of chromosomes at equilibrium. In addition to these dominant chromosomal types there will be an array of chromosomes that arise from these by recombination in each generation, but which are eliminated by selection. For example, in one run with the recombination between adjacent loci of 0·0025, with selection of 10% against homozygotes at each locus, by generation 420 of selection in a population of effective size 400, the following was the observed composition (0s and 1s stand for alternate alleles):

chromosome

001 100 100 011 101 011 111* 100 100 111 000 111	0·440
110 011 011 100 010 100 001* 011 011 000 111 000	0·427
all others	0·133.

Thus all loci are at approximately 0·50 allele frequency except the one marked by an asterisk which is fixed at 1. However, the loci are not associated at random since 87% of all chromosomes are of the two predominant types. Evolution in this population can then *not* be described in terms of allele frequencies alone since the allelic frequencies in this population *have not changed* essentially from their initial values of 0·5. What has changed drastically is the relation between loci, so that a pair of highly correlated structures have been built up.

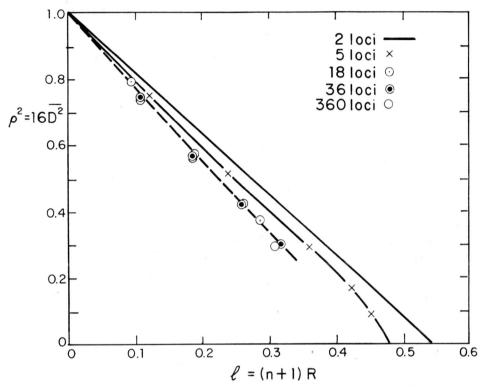

FIGURE 1

▶ *Passage to the limit: elimination of genes.* The most striking feature of the results that Franklin and I have obtained is shown in Figure 1. Here we have taken a given length of chromosome and packed more and more genes into it, dividing the total map length into smaller and smaller intervals. At the same time we have fixed also the total loss of fitness for a completely inbred individual so that, as more and more genes are packed in, the effect of a single gene substitution on fitness gets smaller and smaller. Figure 1 shows what happens when the fitness of a complete homozygote equals 0·0225. On the ordinate is the average squared correlation among all pairs of genes on the chromosome. When $\rho^2 = 0$ the loci are associated completely at random with the frequency of any chromosomal type being given by the product of the allelic frequencies at all the loci. When $\rho^2 = 1 \cdot 00$ there is complete association and there are only two chromosomal types in the population, each at frequency 0·50. On the abscissa is the total map length of the chromosome. Each curve represents the relation between ρ^2 and map length when a different number of genes spans the map length. As we expect, for any given number of genes, as the total map length increases the degree of correlation decreases, since the genes are farther apart. For 2 loci (solid curve) the relation is rectilinear. For 5 loci, the line falls slightly below that for 2 loci and curves slightly at the larger map lengths. Thus breaking up the map into 5 genes instead of 2 has a slight weakening effect on the correlation. It is important to note that in the 2 locus case it is the double homozygote that has a fitness of 0·0225, while in the 5 locus case it is the *pentuple homozygote that has this fitness*. The effect of each gene substitution is much less. For 18 loci again the curve is displaced slightly downward, but for 36 and 360 loci no change occurs! That is, above 18 loci *the relation between degree of correlation and map distance is independent of the number of genes packed into the interval*. Thus if we wish to know how much variation in chromosomal type there will be at equilibrium, we need to know only the total map length and the reduction in fitness in a completely inbred individual (the total inbreeding depression) but we do not need to know how many genes are involved. Indeed, whether separate genes are involved or not is irrelevant and in a way a false question. What gene do we mean? The gene of recombination, the cistron, a cluster of tightly linked cistrons of related function? This problem disappears when we realize that it is the entire chromosome that is the unit of evolution and that it is only the characteristics of the whole chromosome — its map length and its effect on fitness when completely homozygous — that matter in predicting the evolution of the population. Moreover it does not matter that the effect on fitness of a single gene substitution may be vanishingly small, because the linkage distance between adjacent cistrons is also very small and, as Figure 1 shows, in the limit the chromosome behaves as a quasi continuum in which the selection and recombination effects are integrated along the total length of the map.

The results given are not affected either by changing from a multiplicative model to a truncation model of selection, nor are they disturbed by introducing asymmetry. If the two homozygotes at each locus differ in fitness, then the allelic frequencies come to a somewhat different equilibrium, but not a very different one. Indeed, one of the suprises of the asymmetrical fitness models is

R. C. Lewontin

how little different the results are from the symmetric ones. Neither is there any observable effect of varying the map distances between adjacent loci so they are not equally spaced. Apparently the results are of very great generality.

What of our original question about the dimensionality of the evolutionary space? The appearance of highly correlated chromosomal structures with loci not associated at random forced us to say that allelic frequencies at loci were not a sufficient set of dimensions and it appeared that it would be necessary to go to the complete $a^m - 1$ dimensions where every gametic type is an axis of a hyper-tetrahedron. Now, however, we wish to claim that the number of loci is irrelevant and even, in a sense, meaningless. If the chromosome is to be treated as a continuum, then chromosomal 'types' are no longer a denumerable set and we must completely re-orient our theoretical framework. In fact we need a continuous theory of population genetics to deal with the chromosomal continuum. Such a continuous theory will contain only a few variables and a few correlation functions. Allelic frequencies will not appear in its formulations and the notion of the individual locus may disappear entirely from the apparatus of population genetics and evolution. This will be all to the good since we are unable to measure the effect of most single locus substitutions on fitness anyway. In fact it is impossible to tell when two groups of organisms differ at only a single locus, nor do they ever differ by so little in nature. A 'geneless' theory of population genetics will enable us to bring the observables of nature into a rigorous theory for the first time.

References

1. R. C. Lewontin and K. Kojima, The evolutionary dynamics of complex polymorphisms. *Evolution 14* (1960) 458–72.

2. W. F. Bodmer and P. A. Parsons, Linkage and recombination in evolution. *Advances in Genetics 11* (1962) 1–99.

3. M. Kimura, A model of a genetic system which leads to closer linkage by natural selection. *Evolution 10* (1956) 278–87.

4. J. Felsenstein, The effect of linkage on directional selection. *Genetics 52* (1965) 349–63.

5. S. Karlin and M. W. Feldman, Linkage and selection: New equilibrium properties of the two locus symmetric mobility model. *Proc.*

Nat. Acad. Sci. U S. 62 (1969) 70–4.

6. W. Bodmer and J. Felsenstein, Linkage and selection: theoretical analysis of the determinate two locus random mating model. *Genetics 57* (1967) 237–65.

7. M. Kimura, Attainment of a quasi-linkage equilibrium when gene frequencies are changing by natural selection. *Genetics 52* (1965) 875–90.

8. R. C. Lewontin, The interaction of selection and linkage. I. General considerations; heterotic models. *Genetics 49* (1964) 49–67.

9. R. C. Lewontin, The interaction of selection and linkage. II. Optimum models. *Genetics 50* (1964) 757–82.

10. R. C. Lewontin and P. Hull, The

71

On the irrelevance of genes

interaction of selection of linkage. III. Synergestic effect of blocks of genes. *Der Züchter 37* (1967) 93–8.

11. J. L. King, Continuously distributed factors affecting fitness. *Genetics 55* (1967) 483–92.

12. J. A. Sved, T. E. Reed and W. F. Bodmer, The number of balanced polymorphisms that can be maintained in natural population. *Genetics 55* (1967) 469–81.

Complex systems

R. Levins
University of Chicago

The question under consideration is whether we can develop a fairly general and widely applicable theory on the structure and dynamics of complex systems, which would be applicable to work in biology at the level of the population, the cell, development, and perhaps be of some use in the analysis of other complex systems of a social kind, even of the complexity in the evolution of languages and other areas. This does not mean that we can have a theory which will provide the answers to problems in those fields. Obviously the question of what parameters are relevant, what problems are worth looking at, what kinds of answers would be meaningful, is a matter for people working in those fields themselves. However, what we can hope to do with a fairly general theory of complex systems is, as a minimum, to warn against certain kinds of conclusions that do not necessarily follow. In doing this, it is a matter of attempting to over-come the Anglo-American tradition of a reductionist view of things, by very consciously and deliberately examining those properties which follow from complex forms of organization. For example, it has been said by some social scientists that it would be impossible to have a scientific theory of society, because of the intervention of many random factors. In its worst form, by random they simply mean social events which they cannot predict. In its better form, the statement would be that events from outside of social theory, geographical events, meteorological, climatic events, may intervene, or events from societies outside the one they are studying and which are regarded legitimately, therefore, as random perturbations of the system under study. Nonetheless I think we can claim that this argument is wrong. Certainly the study of random processes makes us aware of the fact that there are random processes with determinate outcomes, random processes with several discrete alternatives, and random processes which end up in a random distribution of solutions. The point, there-fore, is not that we have solved the problem for the social scientist, but rather that we have pointed out that his solution is incorrect and that it would not be possible to repair that kind of claim unless it can be demonstrated what kind of random process is involved in the social events under study. So then the first application of a general systems theory approach is the demonstration that some-thing is not necessarily so.

73

Complex systems

We are concerned with two interrelated problems — the ontological issue of how such systems are really put together and the epistemological one of how to study and describe them. Clearly our epistemology must be based on the ontology, but the correspondence between the complexity of the system and the tools for dealing with it is not one-to-one or monotonic. It is true of course that a simple system with few kinds of parts and relations can be dealt with by some simple algorithms. But a system of total complexity, in which many different kinds of elements interact in all directions without constraint, is also simple, The arbitrariness of the structure prevents saying much about it in general, and what we face is second order — or meta — chaos, in which the structural parameters instead of the states of the particles have some probability distribution.

Consider an arbitrary system of N factors, each of which can exist in 2 states. In the absence of special constraints we could design an analysis of variance which relates a measurable trait of the whole to N additive components, $N(N-1)/2$ pairwise interaction coefficients, $\frac{N(N-1)\ (N-2)}{6}$ third order terms, and so on. We would have to estimate 2^N parameters and need a sample of more than 2^N cases. For even a modest N of 10 we need more than 1,024 cases. If we were given a computer with $10^5 \times 10$ digit words in storage, and had to discover its mode of operation by varying the states and observing the output, we would need $(10^5)^{10^5}$ observations, or some $10^{50,000}$ experiments. Clearly, then, a system of this type would be unknowable.

There are several ways around this dilemma. We could look inside. Or, if we knew what a computer was for, we could distinguish program from data, locate the program register, and puzzle it out from there. We could assume that interactions extend only to the third order (instruction, operand, operator). Or we might take a statistical approach, sampling to get estimates of the interaction coefficients of each order, and describe the rough behavior. (But even the sampling approach makes a decency assumption about the distribution of parameters. If some parameter X has a value 0 with probability $1 - \frac{1}{N}$, and N with probability $\frac{1}{N}$, as N increases the mean goes to one but the probability of getting an estimate different from zero goes to zero. Thus we would not detect this interaction at all, but it would have an appreciable effect occasionally as a unique event.)

Therefore, by virtue of Elsasser's principle of finite classes, the totally complex system corresponds to a trivial epistemology which is impotent. Epistemological complexity only arises by constraining ontological complexity : complexity is created by its negation.

But the difficulties of parameter estimation are not the most serious problems in the study of complexity. Suppose that we did know the interrelations among all parts of a system and would describe the rate of change of each variable as a function of the others. Then we would have a very large set of simultaneous non-linear equations in a vast number of variables, and depending on so many parameters, the estimation of each of which may take a lifetime.

These equations will usually be insoluble. They would be likely to be too numerous to compute. If we could compute, the solution would be simply a number. If we could solve the equations the answer would be a complicated expression in the parameters that would have no meaning for us. Therefore the only way to understand a complex system is to study something else instead.

That something else is a model. A model is a theoretical construction, a collection of objects and relations some, but not all, of which correspond to components of the real system. In one sense it is a simplification of nature. We ignore components which we believe to have small effects, or large effects but only rarely. We lump together components which are different. And we leave ambiguities, defining only parts of relation. For instance we might specify that $f(x)$ is an increasing function of x without giving a particular form to $f(x)$. But, in another sense, a model creates complication. We replace the universal but trivial statement 'things are different, interconnected, and changing' with a structure that specifies which things differ in what ways, interact how, change in what directions.

Clearly the choice of model depends on the purpose of the study. I have argued elsewhere that no model can simultaneously optimize generality, realism, and precision. In contemporary population biology the Holling–Watts school derived from applied ecology emphasize realism and precision at the expense of generality. A number of investigators like Edward Kerner and Egbert Leigh have used statistical mechanics approaches to produce models which are general and precise but of dubious realism. And Lewontin, MacArthur, and I have usually stressed generality and to some extent realism at the expense of precision. Since there are no universally optimal models a theory must be a cluster of models which fit together in different ways.

Since the construction of a model is a mapping of components of the real system onto some theoretical space, the isomorphisms of the mapping have been emphasized. Clearly some isomorphism is necessary for relevance, but a

Complex systems

complete isomorphism, a mapping of a system point by point onto itself, would be useless. While the invariants of the mapping preserve relevance, the variable part makes it worth while. Thus, for example, it has been pointed out that the mapping of a flock of sheep onto a set of numbers preserves a measure. But you can take logarithms of numbers, but not of sheep or a flock of sheep. A sequence of flock numbers gives a process, and the flock can now be mapped onto the space of a differential equation, although sheep are not differentiable. Or if the matter of interest is social hierarchy within the flock, ranks are assigned to animals and the flock can be mapped into a permutation group even though a sheep has no inverse.

Models differ in the aspect of reality preserved, in the departures from reality, and in their manipulative possibilities. They can therefore give different results. Since it is not always clear which consequences derive from the properties of nature, it is often necessary to treat the same problem with different models. A theorem is then called robust if it is a consequence of different models, and fragile if it depends on the details of the model itself. The search for robustness leads to the proposition that truth is the intersection of independent lies.

There are many ways in which complex systems might be classified. I would suggest a few, only to indicate a method for approaching this problem. First, consider the aggregate system in which the properties of the whole are statistics of the properties of individual parts, in which the individual parts affect the properties of the whole only by virtue of being part of a mean, or a variance, as a part of frequency, in which the constituent parts therefore are all affecting the properties of the whole in the same way, and are not directly acting on each other in the model. Now, when we say they are not directly acting on each other, this does not mean that there is no physical interacting. Clearly the frequency of genes in males and females affects the population dynamics at the level of the whole population, but we can ignore for conceptual purposes the fact that males and females get together as part of the process of population growth. It is simply irrelevant at the level of the population dynamics for some kind of problems, although not for others.

After the aggregate system, the next kind would be a composed system, such as an engineer's circuit. In this system, the way in which different kinds of parts are strung together into the system will determine system properties. The properties of the system, therefore, are no longer derivable from simple statistics of the components. The different components may be different kinds of units, condensers, transistors, wires, switches, and so forth. Nonetheless this is a composed

76

system because the properties of each component can be completely specified by study in isolation. They do not affect the mode of response of each other, but only the way in which a signal is processed that passes through all of them.

A third kind of system no longer permits this kind of analysis. This is a system in which the component subsystems have evolved together, and are not even obviously separable; in which it may be conceptually difficult to decide what are the really relevant component subsystems. Thus, for example, we might consider that a simple multi-cellular organism is composed of cells, and yet the cells may be more profitably regarded, under other circumstances, as simply spatial subdivisions, partly isolated, of an organism.

The decomposition of a complex system into subsystems can be done in many ways. A plant may be decomposed into leaves and branches and roots and stem, and if we do it this way, leaves are pretty much independent of each other in most of their physiology. On the other hand, a completely different subdivision, into processes of mineral assimilation, photosynthesis, and respiration, will give very different results. What this means is that it is no longer obvious what the proper subsytems are, but these may be processes, or physical subsets, or entities of a different kind. In any case, these main categories of systems, the evolved systems, the composed systems, and the aggegate systems, are obviously sufficiently different so that we must proceed with great caution in attempting to transfer any ideas from one to the other.

In the creative analysis of systems, it becomes crucial to decide what things about the system are worth knowing. There exists a prejudice, perhaps derived, legitimately or illegitimately, from other disciplines, that in order to know something we must define it precisely and measure it precisely. In fact in the development of population biology this has not been the case. The theory of the niche begins as a vague heuristic concept, referring to the fact that organisms are coupled into certain kinds of environment. The subsequent development of the concept involves making some things more precise, changing the definitions, applying it to certain test situations. However, it does not imply that we cannot get anywhere without starting out with a completely formed and rigorous definition of something which is measurable.

It seems that the only way we can understand complex systems is by discarding things that should be known precisely, by deciding on only a few parameters which must be specified, and only specifying these to the degree necessary for the problem to hand. For example, the question of whether genes at a single locus will enter into a stable polymorphism depends on whether the fitness of the homozygotes is less than, or greater than one, which is taken as the fitness of the heterozygote. If we therefore estimate the fitness of the homozygote as 0.99 ± 0.02, we can make very little prediction; either outcome is possible. On the other hand, if the estimate for the fitness of the homozygote is 0.2 ± 0.6, we have a less precise measurement, but a much more precise prediction.

Furthermore I would assert that the goal of prediction is a subordinate one; a contingent objective which is legitimate only to the extent that it helps us decide among alternative hypotheses, or in situations where there are applied problems of importance, but not equivalent to the goal of science itself, which is explanation.

The first task then in the study of complex systems is the identification of sufficient parameters. We have seen that as the model of the organization of the genotype becomes more complex, it becomes increasingly cumbersome to regard the genome as organized in discrete genes and specify their additive effects, their dominance interactions, their additive by additive epistatic interactions, and so forth, and it becomes possible to look at the genome as a whole, defining measures along the whole chromosome. It becomes less interesting to worry about the mutation rate per locus, and more important to know about what proportion of the existing phenotypic variance is due to recent mutation, regardless of how this mutation is spread over different loci. The changes of gene frequency at many different loci simultaneously can be visualised as the movement of a point across a multi-dimensional surface. The roughness of this surface, the extent to which it is broken up by ridges, the extent to which there are distinct saddle points, optima, minima, will determine the overall behavior of the populations, the determinacy of the process, and all the other things that we really want to know. Therefore it would seem that descriptive parameters of this surface, the ridginess, the distribution of peaks, and so on, become more important than the description of the individual genes in the genome, or how the genotype is organized. Similarly in ecological systems we find that the important parameters are niche dimensions; niche breadth, which corresponds to our more vague notion of tolerance for a wide range of environments of the unspecialised condition, and niche overlap, from which we can attempt to derive coefficients of competition.

After we have sufficient parameters for a system, we are faced with the problem of deciding what to do with them. In ecology a major problem is to determine how many species can co-exist. Now we have to do this on the basis of very incomplete knowledge of the forms or the functions expressing

78

competitive interaction. The way we have approached it so far is in a very qualitative fashion. We have specified a certain niche dimensionality. From this, and from the assumption that fitness decreases as the phenotype departs from some optimum value, we can plot the niche of an individual species as a fitness measure on an environment space. We do not know anything about the shape of this fitness measure except for a few special cases. Therefore we have tried a number of them, and we have found that the results are robust; that is, whether we assume that the niche has a rectangular shape, or a Gaussian shape, or a triangular shape, the closeness of packing of co-existent species comes out to be pretty much the same. Therefore, there is a greater confidence in the theoretical results than if we had set up a particular equation and made predictions from this. In the latter case we could never tell whether we were dealing with the consequences of the details of our model or with the natural situation, whether we are really learning something about nature, or we are examining a map through a microscope.

One kind of structuring of systems is a levels model. We do not want to get involved here in the sterile polemics on emergence, or whether things are different despite similarities or similar despite differences. For our purposes a levels model is hierarchical structuring of components into systems and subsystems in such a way that a group of systems are on the same level if the same properties and components are defined for them. A system at a given level has properties defined in terms of the properties of its subsystems and the way they are combined, but the components of a subsystem affect the system only through properties of the subsystem.

The classic example is of course the levels of molecule, cell, organism and population. The dynamics of genetic change in populations depends on the genotypes of the component organisms, but these genotypes appear at the level of population as frequencies and fitnesses.

It is irrelevant to population genetics that fitness depends on complex physiological and behavioral consequences of gene action. All of these factors flow together into the sufficient parameter 'fitness'. But at the organismic level the physiology of survival and reproduction is an object of interest. The genes are relevant only in so far as they result in physiology. The nature of the code and the structure of DNA does not matter at all.

Thus the hierarchical structure of a levels model creates a series of many to few transformations, whereby a large number of parameters at one level flow together into a smaller number of sufficient parameters at the next higher level.

79

Complex systems

The many-to-few nature of the transformation makes it possible to work at higher levels. But it also causes loss of information. If viability, fecundity, and rate of development flow together into fitness, once we know the fitness at the population level we cannot return directly to separate fecundity, viability, and developmental rate. From the parameter fitness on outward we can do population genetics with fitness given. The unraveling of fitness is a separate problem. Thus the hierarchical structure results in the creation of distinct disciplines at each level.

However for all its power in creating manageable models, the levels approach is incomplete. First, there is the problem of proper subsystems. We define a proper subsystem as one whose components contribute to the whole only through subsystem properties. However it often happens that there is an intrusion onto a given level from several levels below. For example, general ecological argument would lead us to expect that as plants adapt to the cold they become more heat sensitive and vice versa. This leads to eco-geographic and temporal predictions. But, according to Jacob Levitt, heat and cold death both proceed by way of the reactivity of -S H bonds, which are brought closer together by dehydration in the case of cold and by thermal agitation in the case of heat. Thus the acquisition of cold resistance enhances heat resistance. This conclusion is not a consequence of any general organismic properties but a molecular intrusion into the affairs of populations.

We have two choices here. We can remove the physical chemistry of protein from the position of component at the organismic level to become a component of the ecological level. But in that case the number of levels from atomic to populational depends on the path along which you count. Level number no longer gives a well-ordered set. The other choice would be to dissolve the organism as a level and let protein chemistries be uniformly the level below population. This grossly increases the number of sufficient parameters. But the decomposition of an improper subsystem to get smaller proper subsystems may go on indefinitely. I assert:

1. A levels model is a hierarchically arranged system of proper subsystems; and
2. There are no non-trivial proper subsystems.

A second difficulty arises because level is not a one-dimensional metric. A local population of a species is a component of its community along with other species. Its important properties are ecological — position in the food web, competitors, prey and predators. It is also a component, along with other local populations of its own species, in the metapopulation linked by gene flow, a

80

pattern of extinction and recolonization, and geographic genetic differentiation. The community and the metapopulation are neither the same level nor is one a component of the other.

Finally, a levels model is ordered, with arrows going from 'lower' to 'higher'. But there are also arrows in the reverse direction. The properties of an organism depend on the cells but the cells are modified by the organism. The behavior of a biochemical network is a consequence of the kinetic constants of the component enzymes. But these enzymes have been selected at the level of the population on the basis of the effect that whole-network properties have on the organism. It is true of course that upward and downward arrows have different meaning.

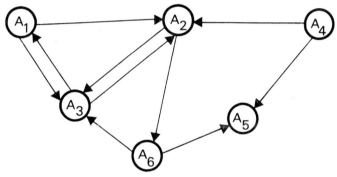

FIGURE 1

Suppose now that we have chosen a particular decomposition into subsystems which we can represent as in Figure 1. The arrows represent the direction of influence. It often happens that processes can be separated into discrete classes by their time constants. Suppose that the rate of change of A_1 is very slow with respect to the other processes. Then it can be regarded as a constant while the other processes evolve. The arrows to A_1 can be suppressed and the arrows from A_1 incorporated into the constants of A_2 and A_3. A_4 does not receive any input from the system. Since it is extrinsically determined, the arrows from A_4 can be replaced by random inputs to A_5 and A_2. Finally if A_6 changes much more rapidly than the other components it can be treated as if it were at the equilibrium value determined by its inputs. Then arrows from A_6 can be joined directly to the arrows into A_6 and A_6 can disappear also from the system, leaving us the reduced structure of Figure 2. This reduced graph defines the system of interest.

81

Complex systems

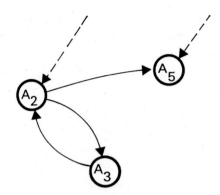

FIGURE 2

It will now be argued that from the graph alone we can discover some of the properties of the system. To do this, we go from a graph to a matrix to the characteristic equation of the matrix which determines the local stability of the system.

Associated with each A_i of the graph is some measure X_i of the state of subsystem A_i. The change in X_i is given by a differential equation

$$\frac{dX_i}{dt} = f_i\,(X_i, X_2, \ldots)$$

where X_j appears in f_i only if there is an arrow from A_j to A_i in the graph. The solutions of the simultaneous equations

$$f_i(X_1, X_2, \ldots) = 0$$

gives the equilibrium points of the system. Now define the matrix of the system

$$A = \begin{vmatrix} a_{11} & a_{12} & a_{13} \cdots \\ a_{21} & a_{22} & \end{vmatrix}$$

where $a_{ij} = \delta f_i / \delta X_j$ evaluated at the equilibrium point. Finally, form the characteristic equation

$$P(\lambda) = \begin{vmatrix} a_{11} - \lambda & a_{12} & a_{13} \\ a_{21} & a_{22} - \lambda & a_{23} \cdots \end{vmatrix} = 0.$$

The system will be stable locally if all the roots λ_i of the characteristic equation have negative real parts; it will be unstable if any root has a positive real part; and it will oscillate inward toward a stable point or outward from an unstable equilibrium if the roots are complex. Therefore the structure of the characteristic

82

equation is an essential property of the system. Several kinds of systems will be examined for qualitative properties.

1. If two systems are not connected at all, their matrix will be

$$M = \begin{vmatrix} M_1 & 0 \\ 0 & M_2 \end{vmatrix}$$

where M_1 and M_2 are the subsystem matrices, and the characteristic equation

$$P(\lambda) = P(\lambda_1)P(\lambda_2).$$

Thus if the characteristic equation of a system is the product of the characteristic equations of its subsystems the systems are structurally separable.

In a hierarchical levels model, let subsystems X_{ji} have parameters y_i which interact in a way described by some graph. Now let each y_i also be influenced by the components X_{ij} of its own subsystem directly, so that the system is graphed as in Figure 3. Then the system can be shown to be separable:

$$P(\lambda) = P_1(\lambda)P_2(\lambda) \ldots P_i(\lambda)P_y(\lambda)$$

where the $P_i(\lambda)$ are the characteristic equations of the subsystems and $P_y(\lambda)$ is the characteristic equations of the Ys. This does not mean that the systems and subsystems are independent. But the components X_{ij} affect $P_y(\lambda)$ only by way of the constants of the matrix a_{ij} which enter the equation. Hence in a hierarchical levels structure the characteristic equations of the different levels are separable.

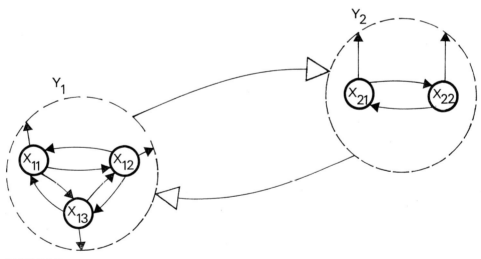

FIGURE 3

Complex systems

Consider now a system of many levels each of which depends on events on its own level and the next lower one. The graph would be an ordered chain and the matrix would have the principle diagonal and the diagonal below it different from zero :

$$M = \begin{vmatrix} a_{11} & & & 0 \\ b_{21} & a_{22} & & \\ & b_{22} & a_{33} & \\ 0 & & & \ddots \end{vmatrix}$$

so that the characteristic equation is $P(\lambda) = \Pi(a_{ii} - \lambda)$ and the characteristic roots are the a_{ii}. Now introduce a single loop down from a higher to a lower level. Then the matrix becomes

$$M = \begin{vmatrix} a_{11} & & & c_{1k} \\ b_{21} & a_{22} & 0 & \\ & b_{32} & a_{33} & \\ 0 & \ddots & & \ddots \end{vmatrix}$$

and the new characteristic equation is

$$P(\lambda) = \prod_i (a_{ii} - \lambda) + (-1)^{k+1} C_{1k} \prod_2^{k-1} (a_{ii} - \lambda) \prod_2^{k-1} b_{i, i-1},$$

or

$$P(\lambda) = \prod_{i > k} (a_{ii} - \lambda) \left\{ \prod_{i=2}^{k-1} (a_{ii} - \lambda) + (-1)^{k-1} C_{1k} \prod_2^{k-1} b_{i, i-1} \right\}.$$

Thus the characteristic equations are still separable outside the loop but confounded within it. Since there is now at least one loop of length greater than two but no loops of length two the system will oscillate (there are complex roots) although the oscillation may be damped. For the special case $a_{ii} = a$, $b_{i, i-1} = b$, the roots from within the loop give

$$P(\lambda) = (a - \lambda)^{k-2} + (-1)^{k-1} C b^{k-2} = 0.$$

This has the roots

$$\lambda = a - C^{1/(k-2)} b \times {}^{(k-2)} (-1)^{\frac{1}{(k-2)}}$$

For large K, $C^{1/(k-2)} \to 1$ and the roots are independent of the magnitude (but not sign) of C.

Suppose now that we have an aggregate system in which all subsystems interact directly, reciprocally, and in the same way only with the system property itself. Then the matrix can be expressed as

$$M = \begin{vmatrix} & & a_{12} & & a_{13} \cdots \\ & a_{21} & \mathbin{|} & & 0 \\ & a_{31} & & \mathbin{|} & \\ & \vdots & & & \\ & \vdots & 0 & \mathbin{|} & \\ & \vdots & & & \mathbin{|} \\ & & & & \mathbin{|} \end{vmatrix}$$

where the first row and column correspond to the system parameter. The characteristic equation is

$$P(\lambda) = (-\lambda)^n - (1 - \lambda)^{n-2} \, \Sigma a_{ij} a_{j1}.$$

Thus the root $\lambda = 1$ occurs $n - 2$ times and the remaining two roots are

$$\lambda = 1 + \sqrt{(\Sigma a_{ij} a_{j1})}.$$

Thus if the interactions are reciprocal so that $a_{1j} a_{j1} > 0$, there is an upper limit to the amount of interaction compatible with stability. The system is stable if $\Sigma a_{1j} a_{j1} < 1$. This will depend on n, on $\bar{a}$, and on the covariance of $a_{1j} a_{j1}$. The latter may be a function of the variance of the steady state values of X_i.

Therefore if the differential equations for the variables are non-linear so that alternative equilibria are possible we could look at the statistical distribution of the equilibrium values X_i. The matrix may tell us that only if the X_i are less variable (or more variable) than some level can the system be stable.

We now return to the original question : in a world with so many components and so much potential interaction, why is it possible to have any regularity, why is it possible to isolate any subsystems, why is anything simple?

A number of quite independent lines of argument converge toward the assertion that there is often a limit to the complexity of systems.

First, there are optimality arguments. Kauffman's random nets seem to give

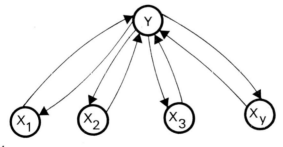

FIGURE 4

viable results only for low connectivity, and then almost any realization already exhibits complex behavior. I have argued that when conflicting fitness criteria are imposed on a large number of genes in a developing system, subsystems will become uncoupled as a result of natural selection and we will end up with clusters of genes, each evolving under the control of relatively few fitness criteria. Ashby, in his design for a brain, argues for loosely coupled subsystems as optimal.

Secondly, there are time considerations. A random net with very high connectivity will take an immensely long time to go through a cycle. But selection is a statistical phenomenon. A very long cycle has no real average performance, and could not be selected. Further, a genetic system with very strong epistatic interactions would have a very rough 'adaptive landscape', and Boscert has shown why such systems would respond very slowly to optimization by selection.

We now introduce one final line of argument. The dynamics of a broad class of complex systems will result in simplification through instability. That is, some of the interacting components may drop out and each remaining component will interact strongly with a few others only. In that case each component is partly autonomous : its direct effect on its own dynamics ($\delta f_i / \delta x_i$) is a significant fraction of all influences received ($\sum_j \delta f_i / \delta x_j$) and greater than its indirect

effect on itself by way of other components $\left(\sum \dfrac{\delta f_i}{\delta x_j} \dfrac{\delta f_j}{\delta x_i} \right)$. The pattern in which

components drop out may produce clusters, hierarchies, and so on, while the effect of a component on itself may be the result of its own interacting subsystems.

Alternatively, very strongly interacting components can lose their identity in a field. Then the system is better approached in terms of waves rather than particulate components.

The model we use is that of complex systems with global structural homogeneity. We define a system with global structural homogeneity as follows.

1. Let A be a space for which a measure of distance exists, and let the A_i be elements in this space. In an ecological space the axes represent environmental components, the elements are species, the location of an element defines its niche, and the distance between elements is an inverse measure of niche overlap or competition. In a biochemical space the (ordered) axes may stand for the amino acids in a polypeptide chain, and so on. In a model of the nervous system,

distance may stand for number of synapses between a pair of neurons. Then a hierarchically structured nerve net would be represented by a space in which 'volume' increases exponentially with radius instead of as an integral power.

2. The structure of A within a radius r about any element may depend on r but not on the choice of element.

3. To each A there is a quantitative measure x_i such that

$$\frac{dx_i}{dt} = f_i(x_1, x_2, \ldots).$$

Each f_i is a permutation of the variables x_j. Thus $\dfrac{\delta f_i}{\delta x_j}$ is a monotonic decreasing function $a(r)$ of the distance between A_i and A_j.

The matrix of the system of equations is made of elements $|a_{ij}| = a(r)$. The a_{ij} may depend on the variables but for the simple ecological case they do not.

We will use the following results, which are asserted without proof.

1. The system whose stability matrix has diagonal elements $a_{ii} = 1$ and general elements a_{ij} is stable if $\sum\limits_j |a_{ij}| < 1$ for all i.

2. If for matrices of order $n > n_o$, $\sum\limits_j a_{ij}a_{ij} > 1$, then there is an n_1 such that the system is unstable for $n > n_1$. That is, there is an upper limit to the size of such systems compatible with stability.

The use we make of stability or instability depends on the system. In the ecological community, species will disappear from unstable communities. But for systems bounded away from 0 and with some upper bound, instability implies spontaneous activity – often oscillation.

The next step is to calculate $\sum\limits_j |a_{ij}|$ and $\sum\limits_j a_{ij}a_{ij}$. Let $V(r)$ be the surface area of a hypersphere of radius r in A. Let C be the density of elements in the space, so that there are $CV(r)$ elements at distance r. Then

$$\sum_j |a_{ij}| = C\sum_r V(r) \, a(r)$$

and $\quad \sum\limits_j a_{ij}a_{ji} = C\Sigma V(r) \, a^2(r).$

We note briefly two cases.

Case 1. $\overset{\infty}{\Sigma}V(r) \, a^2(r)$ diverges. Then $\Sigma V(r)a(r)$ also diverges. This means that the effects of other elements on a given element come mostly from a distance; there is no local neighborhood of radius r_o which can be isolated to study changes in x_i; the total influences acting on two nearby elements are almost identical, and instead of a discrete system we have a field. The field will be unstable unless below a certain size.

Case 2. $\overset{\infty}{\Sigma} V(r)a(r)$ converges. Then for all C we can find some r_1 such that
$C \overset{r_1}{\underset{r_0}{\Sigma}} V(r)a(r) < 1$, and a small densely packed system can be stable. Further,
there is always some C_1 such that $C_1 \overset{\infty}{\underset{r_0}{\Sigma}} V(r)a(r) < 1$, giving a stable system
of infinite extent but loosely packed. There is also some C_2 such that
$C_2 \overset{\infty}{\Sigma} V(r)a^2(r) > 1$ so that a densely packed system will become unstable
if too big. This result explains why there are discrete species instead of a
continuum of organic forms. It could also be extended to problems such as why
there cannot be too many too similar molecular species in cells, why neurons
cannot be directly connected to too many other neurons, and so on. Further
work with this simple model shows the role of the dimension of the space and
the origin of hierarchical structuring ; allows asymmetric interactions ; replaces
interaction coefficients with probabilities of interactions; and so on.

The study of global structural homogeneity is a useful approach to complex
systems for which we can define a distance measure such that the strength of
symmetrical interactions decreases with distance, or the probability of interactions decreases with distance, or for which either holds for distances beyond
some threshold distance r_t. Even this superficial description of some preliminary
results indicates:

1. A weakly specified and structurally simple system can generate rich spatial
and temporal properties if the space is either large enough or dense enough.
2. There is an upper limit to the amount of reciprocal interactions a stable
system can tolerate – things cannot be too interconnected or complex. A tightly
packed, large space can have the following consequences : (a) loss of local
determination and individuality of elements into a field; (b) uniform reduction
of density, producing (c) restriction of the size of the space, producing local
dense clusters of elements widely separated from other such clusters. This is one
form of hierarchical structure. In the ecological case, the result is the reality of
taxonomy of higher categories. In organisms, it suggests that atoms interact very
strongly only with relatively few others on the same molecule; much more
weakly with the many more in the other molecules of the cell; cell to cell
reactions are, on an energy/unit mass basis, weaker yet, and so on. This kind
of clustering also leads to separation of time scales of reaction and hence to
separable levels in the sense of the previous section.
3. Thus, in a model which initially allows unlimited complexity a secondary
simplicity arises, but in the context of an already given complex whole.

Topological models in biology

René Thom

Institut des Hautes Études, Bures-sur-Yvette

Introduction. The problem of morphogenesis – broadly understood as the origin and evolution of biological structures – is one of the outstanding questions in present day biology. Many experimental attempts have been made to elucidate the causes of morphogenetic processes in embryology, development, regeneration, and so on. Some of them have been partially successful. For instance, as a typical example, let us consider the well-known fact of orientation of a plant toward light (positive phototropism); here, the physiologists have been able to characterize a chemical substance, *an auxin*, which inhibits the growth of the stem when under light. In such a case, the immediate causative agent and a satisfactory local explanation have been found. But, in most cases, when one tries to get beyond the first causative factor, the experimentalist gets lost in the seemingly infinite multiplicity of possible causes, and the bewildering variety of intermingled reactions which have to be considered. Most people – in this situation – satisfy themselves by vague appeals to differential action of genes, decoding of genic DNA, and so on.

There is little doubt, in fact, that the problem is essentially of a theoretical, conceptual nature. Granted that all local morphological or physiological pheno-mena inside a living being occur according to a local biochemical determinism, the problem is to explain the stability and the reproduction of the global spatio–temporal structure *in terms of the organization of the structure itself*. There appears to be a striking analogy between this fundamental problem of theoretical biology and the main problem considered by the mathematical theory of topology, which is to reconstruct a global form, a topological space, out of all its local properties. More precisely, a new mathematical theory, the theory of *structural stability* – inspired from qualitative dynamics and differential topology – seems to offer far-reaching possibilities to attack the problem of the stability of self reproducing structures, like living beings. But – at least in the author's opinion – the validity of this type of dynamics description exceeds by far the bio-logical realm, and may be applied to all morphological processes – whether animate or inanimate – where discontinuities prohibit the use of classical quan-titative models. It should be noted, in that respect, that any morphological process involves by definition some discontinuity of the phenomenological

89

properties of the medium studied : this explains why morphogenesis — whether in biology, as in development — or in inanimate nature, as for crystal growth — has up to now resisted all attempts of classical mathematical treatment : any quantitative model using explicit equations involves necessarily analytic, hence continuous functions. The only — partial — exception to this statement is the theory of shock waves in fluid dynamics, where some local equations of propagation may be established, but, here again, complicated problems like the behavior of interacting shock waves may be solved only empirically [1]. In all these situations a new mathematical theory nearer to the qualitative thinking of the topologist than the quantitative estimates of classical analysis seems particularly relevant.

THE MATHEMATICAL THEORY OF STRUCTURAL STABILITY

▶ *Notion of dynamical system.* Suppose we put, in a box B, k chemical substances, $s_1, s_2, \ldots, s_k$, at concentrations $x_1, x_2, \ldots, x_k$. Because of the reactions taking place between these substances, their concentrations x_i vary according to a law which we may write :

$$dx_i/dt = \mathbf{X}_i\,(x_j, \tau, t) \tag{1}$$

where t denotes time, τ some external parameter like temperature. In such a case, a state of the system is described by a system of $(k+2)$ parameters (x_i, τ, t), that is by a point in $(k+2)$-dimensional Euclidean space $\mathscr{R}^{k+2}$, which is the 'phase space' M of our system. The right hand side $\mathbf{X}_i$ of eq. (1) defines in M a vector field $\mathbf{X}$. Provided this vector field satisfies some regularity conditions (for instance to be differentiable), then we may — at least locally — integrate the differential system of eq. (1) and get equations :

$$x_i = h_i(x_i^o, \tau, t) \tag{2}$$

describing the evolution of the system as a function of the initial data x_i^o. This general picture applies to practically all known systems of any nature whatsoever, provided they are directed by a local determinism. The most outstanding example of this model has been given by celestial mechanics, with Newton's gravitation law defining the right hand side of eq. (1) in the phase space (q_i, p_i) of positions and momenta. The differential model $(M, \mathbf{X})$ offers the ultimate motivation for the introduction of quantitative models in science. Nevertheless, its use is fraught with grave difficulties.

1. Despite the widespread belief to the contrary, there are very few natural phenomena which allow of a precise mathematical description, for which the right hand side of eq. (1) is 'exactly' known and given by explicit formulae.

R. Thom

Gravitation and classical electromagnetism are practically the only cases to fulfil this requirement. In most other cases, the right hand side of eq. (1) is known only approximately through empirical formulae.

2. Even if the right hand side of eq. (1) is given explicitly, it is nevertheless impossible to integrate formally eq. (1). To get the solution in eq. (2), one has to use approximating procedures.

For these two reasons, one has to know to what extent a slight perturbation of the right hand side of eq. (1) may affect the global behavior of the solutions in eq. (2). To overcome — at least partially — these difficulties, the mathematician Henri Poincaré introduced in 1881 a radically new approach, the theory of 'Qualitative Dynamics' [2]. Instead of trying to get explicit solutions of eq. (1), one aims for a global geometrical picture of the system of trajectories (eq. 2) defined by the field X. If this can be done, one is able to describe qualitatively the asymptotic behavior of any solution. This is in fact what really matters : in most practical situations, one is interested, not in a quantitative result, but in the qualitative outcome of the evolution (Will the bridge stay or break down?). Thus, qualitative dynamics, despite the considerable weakening of its program, remains a very useful — although very difficult — theory.

▶ *Structurally stable dynamical systems.* A new development occurred in 1935 with the introduction by the Soviet mathematicians Andronov and Pontrjagin of the concept of *structurally stable dynamical system.* The dynamical system (M, X) is said to be *structurally stable* if, for a sufficiently small perturbation δX of the vector field X, the perturbed system $(M, X+\delta X)$ is, roughly speaking, topologically isomorphic to the unperturbed system (See Figures 1 and 2 for examples). An outstanding question was then to characterize the structurally stable systems for a given space M, and to know if they are *dense*, that is, if any differential system in M can be approximated by a structurally stable one. These difficult questions were recently solved affirmatively for dim $M \leq 2$ by Peixoto [2] ; for dim $M \geq 4$, Smale showed that structurally stable systems may not be dense [3].

This generally negative answer to the problem of the density of structurally stable systems shows only that the notion of structurally stable systems is still too fine to be really useful. A further weakening of the notion is obtained by the following consideration : as already remarked, the most important feature of a solution (eq. 2) of eq. (1) is its asymptotic behavior for t tending to $+\infty$. It may happen, for instance, that the representative point $h(t)$ tends toward a point q, which is an *equilibrium position* of the system. If this point q is such

91

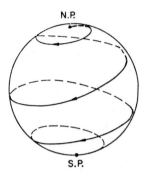

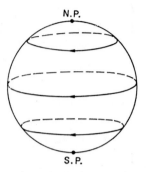

FIGURE 1
Example of a structurally stable system on the
two-sphere $x^2+y^2+z^2 = 1$.

FIGURE 2
On the same sphere, the vector field defined
by $dz = 0$ is not structurally stable; a small
deformation of the form $z = -e^2$ transforms
it into the field of Figure 1.

that the trajectory of any point near q goes to q, and no trajectory leaves q, we
shall say that q is an *attractor* of the system (stable equilibrium). This attractor
is said to be *structurally stable* if any perturbation – sufficiently small – of the
given system contains an attractor q' near to q. For some vector fields, like the
gradient fields, almost any trajectory goes to an attractor – in general a point
which is structurally stable. One may conjecture that for almost any field on a
space M, almost any trajectory goes to an attractor – which may be a more
complicated geometric object, like a closed trajectory, a torus, or an even more
complicated set, but which is nevertheless structurally stable. We might consider,
finally, only those systems which have a finite set of structurally stable attractors.
There are good reasons [3] to believe that in the end this idea will be most
useful, and that any system may be approximated by one of this standard type.
For any such system, let $A_1, A_2, \ldots, A_r$ be its attractors. To any attractor A_i we
associate the set $B(A_i)$ of trajectories tending to A_i, the *basin* of A_i. Almost all
of the space M is partitioned into the basins $B(A_i)$, and the geometry of these
basins characterize entirely the qualitative behavior of the system [4]. In the
simplest cases, like the gradient fields, the basins are separated by piecewise
differentiable hyper-surfaces (like the crest line in a geographic map separating
the basins of two rivers) and these separatrices are structurally stable (see
Figure 3); but, in other situations, the basins may become intermingled in a
very complicated structurally unstable way. If so, the final evolution starting
from a point adherent to these basins may be practically indeterminate, and this
in a 'structurally stable' way. See Figure 4. This shows the philosophically

R. Thom

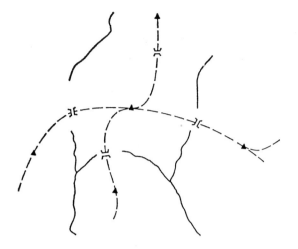

FIGURE 3
The crest line (divide), dashed, separating the basins of two rivers is a piecewise differential curve (having eventually cusps at the generic vertices (summits)).

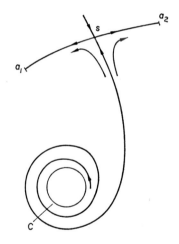

FIGURE 4
Example of intermingled basins: the separatrix arriving at a saddle point s tends to limit cycle C. Any point of C is adherent to the two basins of the attractors a_1, a_2 connected to s.

93

important fact that a deterministic system may exhibit, in a 'structurally stable way', a complete indeterminacy in the qualitative prediction of the final outcome of its evolution [5]. In such a case, we may speak about a 'choice' of the system between the two outcomes, or of a conflict, of a 'fight' between the two attractors. We will return to this point later.

▶ *Structurally stable mappings.* In many cases, the description of a physical process by a dynamical system $(M, \mathbf{X})$ is unnecessarily complicated, and we may — at least locally — parametrize the states of the system by a set of mappings $U \xrightarrow{g} W$ (U, W, Euclidean spaces), which we may suppose to be differentiable.

For instance, if the vector field $\mathbf{X}$ is a gradient field, we may consider instead of $\mathbf{X}$ the associated potential function V ; $\mathbf{X} = -$ grad V, where V is a real, valued function on M ; $V : M \to \mathscr{R}$. Suppose we perturb the given mapping g. We may ask whether the perturbed mapping has the same form, the same 'topological type' as the initial mapping. This gives rise to the problem of stability of differentiable mappings, which is the object of current work among mathematicians [6]. We shall discuss here a special case of the problem, which seems to offer many applications : this is the case of an isolated singularity of a potential function V.

First, let us recall that a singular point of a differentiable, real, valued function V of n variables, $x_1, x_2, \ldots, x_n$, is a point where all partial derivatives of first order $\delta V / \delta x_i$ vanish. (For the dynamical system defined by $X = -$ grad V, these points are equilibrium positions of the system.) The first question is : when is such a singular point structurally stable? Mathematically, the answer is quite easy : a singular point m_o of V is structurally stable if and only if the rank of the mapping defined by $(x_1, x_2, \ldots, x_n) \to \delta V / \delta x_1, \delta V / \delta x_2, \ldots,$ $\delta V / \delta x_n$ is equal to n at m_o, that is, if the hessian $|\delta^2 V / \delta x_i \delta x_j|$ does not vanish. In such a case the point is said to be a *non-degenerate critical quadratic point,* which means that the quadratic part in the Taylor expansion of V at m_o is non-degenerate. Around such a point, there exists a local system of (curvilinear) coordinates, in which V is expressed as a quadratic form [7]

$$\sum_{i=1}^{n} (\pm) X_i^2 = V - V(m_o).$$

If the singular point m_o is non-structurally stable, two cases may occur : let us perturb V by adding an arbitrary function δV such that it and all its derivatives $\partial_\omega \partial V / \partial x_\omega$ of any order are small. Either we may get an infinite number of topological types for the germ of the perturbed function $V + \delta V$; or we can get only a finite number of them. In the first case, the singular point is said to be of *infinite co-dimension,* in the second, of *finite co-dimension.* For instance, in one

94

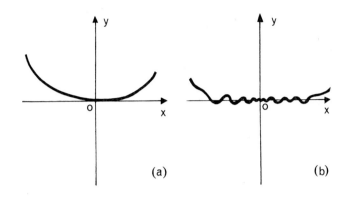

(a)   (b)

FIGURE 5
(a) $y = \exp\left(-\frac{1}{x^2}\right)$ (b) $y = \exp\left(-\frac{1}{x^2}\right) + \cos nx \ \mathrm{esp}\left(\frac{-1}{x}\right)$.

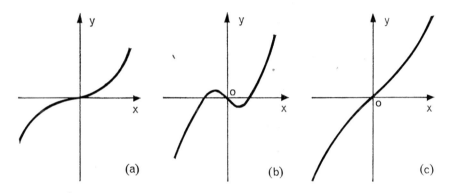

(a)   (b)   (c)

FIGURE 6
(a) $y = x^3$ (b) $y = x^3 - x$ (c) $y = x^3 + x$.

variable, the 'flat' singular point $V = \exp(-1/x^2)$ is of infinite co-dimension, as one may approximate it by a function presenting an arbitrary high number of bumps (take for instance $\delta V = \exp(-1/x) \cos nx, x > 0$). The singular point $V = x^3$, on the other hand, is of finite co-dimension (one), as any perturbation of x^3 is either of the topological type of $x^3 - x$ (curve with a bump), or $x^3 + x$ (curve without bump) (compare Figures 5 and 6). There exists a precise algebraic criterion which tells whether or not a singular point is of finite co-dimension [8].

▶ *Universal unfolding of a singularity.* For a singular point of finite co-dimension, $V(x)$ at $x = 0$, there exists a k dimensional family of deformations which is in some sense universal with respect to all possible deformations of V : If

95

$$V = V(x) + u_1 \, g_1(x) + u_2 \, g_2(x) + \ldots u_k \, g_k(x), \quad u_j \, e \, \mathscr{R},$$

is this universal family, any perturbation of V, of the form $G(x, t)$, with $te\mathscr{R}$ and $G(x, 0) = V(x)$, may be obtained, up to topological equivalence, by a mapping $t \rightarrow u$ in this universal family. For instance, if $V = x^3$, its universal unfolding family is $V = x^3 + u.x$.

This universal family is called – for obvious intuitive reasons – the *universal unfolding* of the singularity $V(x)$, which we call – by analogy borrowed from embryological induction – the *organizing center* of the family. The dimension k of the universal unfolding is the *co-dimension* of the singularity $V(x)$.

This notion of universal unfolding plays a central role in our biological models. To some extent, it replaces the vague and misused term of 'information', so frequently found in the writings of geneticists and molecular biologists. The 'information' symbolized by the degenerate singularity $V(x)$ is 'transcribed', 'decoded', or 'unfolded' into the morphology appearing in the space (u) of external variables which span the universal unfolding family of the singularity $V(x)$.

It is not too difficult a task to find all possible singularities $V(x)$ of finite co-dimension not exceeding four. These singularities are important, because they may appear on our space–time in a structurally stable way. They give rise to what we call the 'elementary catastrophes', when we interpret them as describing dynamical fields on our space–time, as explained below. A list of these singularities, with their dynamical interpretation in everyday language is shown in Table 1.

These 'elementary catastrophes' describe also the structurally stable singularities presented by *wavefronts*, or more generally any propagative process directed by a variational principle, like Fermat's principle in classical optics. This is why it is possible to realize them as singularities of caustics of light rays (see Plates 1–4).

DETERMINISM AND STRUCTURAL STABILITY

The hypothesis that external phenomena are subjected to a rigid determinism is more an epistemological postulate than a proven fact. Not only because of quantic indeterminism : many situations in the macroscopic world exhibit a kind of very high instability : an infinitesimal change of the initial data may cause an enormous change in the following evolution. As a typical example, let us consider the 'bathtub experiment' : in a perfectly cylindrical bathtub, filled with water perfectly at rest, what occurs if we open the plug at the bottom ? Water

TABLE 1

Table of ordinary catastrophes on four-dimensional space–time

Co-dimension	Name	Organizing Center	Universal Unfolding	Spatial Interpretation	Temporal Interpretation
Dimension one					
0	Simple Minimum	$V=x^2$	$V=x^2$	A being An object	To be To last
1	The Fold See Figure 7	$V=x^3/3$	$V=x^3/3+ux$	The boundary The end	To end To start
2	The Cusp (Riemann–Hugoniot catastrophe) See Plate 1, Figures 8 & 9	$V=x^4/4$	$V=x^4/4+ux^2/2+vx$	A pleat A fault	To separate To unite To capture To generate To change
3	The Swallow's Tail See Plate 2, Figure 10	$V=x^5/5$	$V=x^5/5+ux^3/3+vx^2/2+wx$	A split A furrow	To split To tear To saw
4	The Butterfly See Figure 11	$V=x^6/6$	$V=x^6/6+tx^4/4+ux^3/3+vx^2/2+wx$	A flake A pocket A scale (of a fish)	To fill $\Big\}$ (a pocket) To empty To give To receive
Dimension two					
3	The hyperbolic Umbilic See Plate 3, Figure 13	$V=x^3+y^3$	$V=x^3+y^3+w\,xy-ux-vy$	The crest (of a wave) The arch	To break (for a wave) To collapse To engulf
3	The elliptic Umbilic See Figure 14	$V=x^3-3xy^2$	$V=x^3-3xy^2+y(x^2+y^2)-ux-vy$	The needle The spike The hair	To drill $\Big\}$ (a hole) To fill To prick
4	The parabolic Umbilic See Plate 4, Figure 15	$V=x^2y+y^4$	$V=x^2y+y^4+wx^2+ty^2-ux-vy$	The jet (of water) The mushroom The mouth	To break (for a jet) To open $\Big\}$ (the mouth) To close To pierce To cut, to pinch To take, to eject To throw

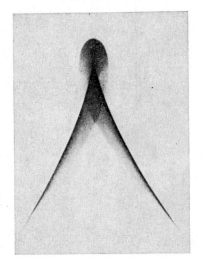

PLATE 1 Cuspidal caustic. (Riemann–
Hugoniot catastrophe.)

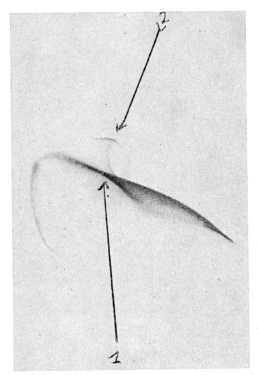

PLATE 2 A swallow's tail (arrow 1). A hyperbolic
umbilic may be seen at the extremity of arrow 2.

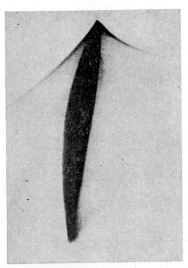

PLATE 3
Hyperbolic umbilic.

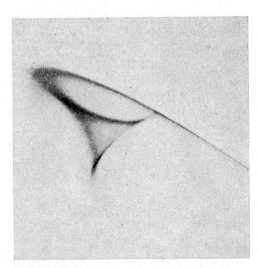

PLATE 4
A deformation of a parabolic umbilic.

PLATE 5
The hydraulic model of the epigenetic landscape.

PLATE 6
The hydraulic model of reproduction. (The clay models were kindly built by M. Marcel Froissart.)

Topological models in biology

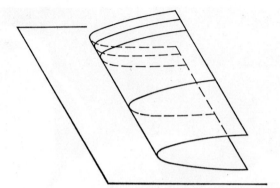

FIGURE 7
Ordinary fold.

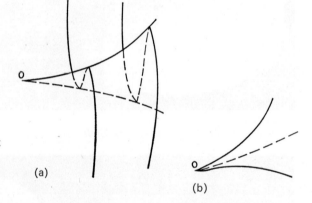

FIGURE 8
(a) Cusp or Riemann–Hugoniot
catastrophe. (b) The universal
catastrophe set of the
Riemann–Hugoniot type (shock
wave with free edge).

(a)

(b)

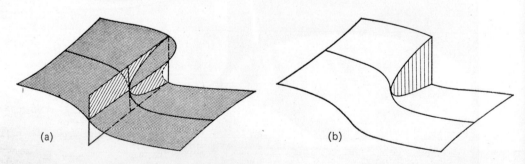

(a)

(b)

FIGURE 9   Riemann–Hugoniot catastrophe

100

R. Thom

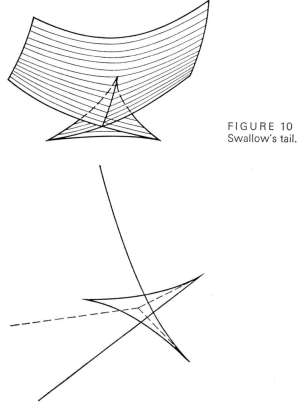

FIGURE 10
Swallow's tail.

FIGURE 11
Most complicated plane section of the universal unfolding of the 'Butterfly' singularity.
Dotted line, the universal catastrophe set.

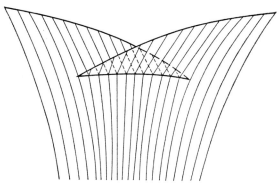

FIGURE 12
Universal catastrophe set associated to the 'Butterfly' singularity: exfoliation of a shock wave.

101

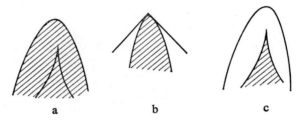

FIGURE 13
Sections of the universal unfolding of the hyperbolic umbilic. Hatched ; domain of a stable regime.

FIGURE 14
Universal unfolding of the elliptic umbilic. Dotted line ; limiting surface of an unstable regime.

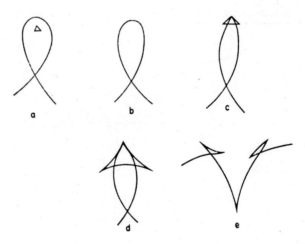

FIGURE 15
Some plane sections of the universal unfolding of the parabolic umbilic.

102

begins to spin in a cyclic movement, the sense of which is practically indeterminate. (Many factors have been invoked to determine it : residual motion
of water; movement of the air at the surface; Coriolis effect due to the earth's
rotation; and so on.) In such a situation, the final state exhibits less symmetry
than the initial data : a *breaking of symmetry* does occur. It is quite clear, for
a priori reasons, that any phenomenon exhibiting such a breaking of symmetry
cannot be given any deterministic, formalisable model. One may still say that the
initial data were not perfectly symmetrical; but proving the deterministic character of the process may well be an *undecidable* question, just as the non-con-
tradiction of arithmetic is undecidable.

Contrary to these highly unstable situations, there are many cases where the
determined character of a process is experimentally obvious; this occurs for
processes described by 'well-posed problems' in analysis, where the evolution is
a continuous function of the initial data; this may occur also for morphogenetic processes, described by a set of discontinuities in the properties of the
medium. If such is the case, if the given morphology exhibits stability properties
with respect to small variations of the initial data, we shall say that the process
is the support of a *morphogenetic field* or, to use a word coined by C. H. Waddington [9], that the process is described by a *chreod*. With this definition,
there is no mystery at all in the notion of morphogenetic field : it only expresses
the fact that a given process gives rise to a fixed morphology — defined once
and for all by a model associated to the field — and this in a structurally stable
way. This definition may be put in a precise mathematical form involving the
topological notion of homeomorphism.

For any natural morphological process, it is very important to isolate first those
parts of the process which are the support of morphogenetic fields to find out
the *chreods* of the process. They form kind of islands of determinism, separated
by zones of instability or indeterminacy. That such a presentation is possible
amounts to saying that the morphology is more or less *describable*. In fact,
almost any natural process exhibits some kind of local regularity in its morphology, which allows one to distinguish recurrent identifiable elements denominated
by words. Otherwise the process would be entirely chaotic, and there would be
nothing to talk about. (*Turbulence* in hydrodynamics might be an example of
the last kind, and one knows the difficulty met just in *describing* the process.)
▶ *Semantic models*. This decomposition of a morpohological process taking
place on an Euclidean space $\mathscr{R}^m$ can be considered as *a kind of generalized
m-dimensional language*; I propose to call it a *'semantic model'*.

Topological models in biology

In fact, our usual language is nothing but a *semantic model of dimension one* (the time), the *chreods* of which are the *words* (spoken, or written).

Given such a 'semantic model', then two kinds of problems may be considered:
1. To classify all types of chreods, and to understand the nature of the dynamic processes which insure their stability.
2. A process involving several chreods is in itself structurally unstable (otherwise it would be covered by a unique over-chreod); but, frequently, one has to deal with an *ensemble* of processes of the type studied. Then, generally, there are some associations of chreods which appear more frequently than others. One may speak, in that case, of a *multi-dimensional syntax* directing the semantic model. The problem is then to describe this syntax, to formalize it in the same way as one may formalize grammatical rules in linguistics.

To do that, one needs to build, first, a dictionary of chreods, second, what the linguists call a 'corpus' of the given language; it is the task of the experimentalist to give this 'corpus' in the case of natural morphological processes, and to extract from it statistical data. This is in fact what quantitative biologists do in forming statistics of morphological processes; physicists do the same in their scattering experiments in elementary particle physics. The problem of interpreting these data, and extracting out of them a formal theory, seems to be – in general – of the utmost difficulty; it amounts to deciphering an entirely unknown language.

Going back to the first problem, what would be its interpretation in usual linguistics? This would be the famous problem of Plato's Cratylus, to understand how the phonetic structure of a word proceeds from its meaning. One knows that, in that case, the relation between the structure of a word and its meaning is very remote, darkened as it is by the effect of a long history. In many natural phenomena – especially of inanimate nature – such an arbitrary coding is not to be expected, and one might hope to read more or less directly from the internal structure of the chreod the qualitative dynamic which insures its stability. In biology the situation is somewhat intermediate : in some cases, the dynamical interpretation of a morphological process is fairly easy; in other cases, the weight of the past manifests itself by submitting the process to *genetic constraints*, which makes the dynamical interpretation more difficult and sophisticated.

THE GENERAL DYNAMICAL MODEL

Bifurcation and catastrophes. Let U be a domain in space–time in which some natural process takes place; we admit that all possible local states of the process can be parametrized by points of a manifold M, and that the local evolution

104

around a point $n \epsilon U$ is described by a vector field $\mathbf{X}(u)$ in M, varying slowly with u. Then, the local dynamic around u reaches a structurally stable attractor (a stable regime), and stays there for u varying in U, until we reach a point in u when this attractor breaks down through the variation of the dynamical system $X(u)$; the final state is then captured by another attractor (the 'basin' of which is adherent to the vanishing basin of the destroyed attractor). We get in U a 'shock wave' separating the two regimes, which defines in U the morphology to be studied.

This shows that, in such a model, the fundamental phenomenon to be studied is the destruction of a structurally stable attractor by variation of the vector field. This is the object of a part of qualitative dynamics named — after Henri Poincaré — *bifurcation theory*; this theory is far from being well known from a mathematical point of view. The morphological effect of such a change in local regime I propose to call a *catastrophe*. Our main postulate is that any morphology can be attributed to such a bifurcation phenomenon, whatever may be the nature of the ambient medium and the physical nature of the forces acting in the local dynamic. Explanations of this kind (with the local metabolism as local dynamics) were put forward for cellular differentiation by C.H.Waddington and Max Delbrück around 1940 [10]. But the great forerunner in this field of ideas is d'Arcy Thompson [11], whose famous treatise *On Growth and Form* contains a wealth of examples and ideas which still have to be explored and developed from the mathematician's point of view.

▶ *Catastrophes and morphogenetic fields.* It remains now to explain how bifurcation theory of dynamical systems may lead to the notion of *morphogenetic field* of a *chreod*. Here, the intuitive notion is that even *bifurcation and catastrophe may occur in a structurally stable way, according to a fixed algebraic model* given by theoretical considerations. This is true at least of the most simple type of catastrophes — which we call *ordinary catastrophes* — by contrast to *generalized catastrophes*, to be described later.

In the actual state of the mathematical theory, the study of bifurcation — and the ensuing catastrophes — can be done only when the local dynamic $(M, \mathbf{X})$ is a gradient dynamic. In that case, the theory of bifurcation reduces to the theory of structurally stable mappings, and the notion of the 'universal unfolding' of a singularity of the potential may be applied. On the universal unfolding space (spanned by the external variables u_i) we get a system of shock waves describing the conflict between the attractors (minima of the potential of the internal dynamic $(M, \mathbf{X})$). One gets this system by applying the somewhat arbitrary —

but easy — rule known as 'Maxwell's convention' : On any point u of the un-folding space U, the dominating regime corresponds to the absolute *minimum of the potential V*. By this rule, we may associate to any singularity of the potential $V(x)$ of finite co-dimension a *universal catastrophe set, K,* defining in the unfolding space U with the singularity $V(x)$ as *organizing center*. If in some domain W of space—time $\mathscr{R}^4$, the local internal dynamic has the singularity $V(x)$ at a point $w \epsilon W$, then — in general — the associated morphology is given by a mapping h of the domain W into the unfolding space U, and we may suppose this map to be in general position transversal to the universal catas-trophe set K. Then the morphology which appears around w in W as a result of the bifurcation $V(x)$ is the counter-image $h^{-1}(K)$. Roughly speaking, the morphology having $V(x)$ as organizing center is given by the universal model K; there exists around w a *morphogenetic field*, a *chreod* which describes the induced morphology. As the set K has a relatively simple topology — a polyhedral structure — the induced morphology is itself relatively simple : this is the case of *ordinary catastrophes*.

In order that such a process takes place, a preliminary condition has to be satisfied : the external variables u_i of the unfolding space must have some local realization as coordinates in W. This requires that the domain W be *polarized* by local agents. The necessity of a preliminary polarization in a tissue support of a morphogenetic field is a postulate of *Child's gradient theory* in embryology [12]. Our model justifies entirely this point of view.

When the domain W is not sufficiently polarized, then the mapping h may not be transversal to K; then, the induced morphology $h^{-1}(K)$ may be quite com-plicated : we get what I call a *generalized catastrophe*. A generalized catas-trophe is characterized by a very complicated topology involving ramifying domains into smaller and smaller pieces (or, conversely, the condensation into isolated clumps of a dust of very fine particles). Such generalized catastrophes appear as a rule in all symmetry-breaking processes, and they are in general structurally unstable (although the final state of the catastrophe may be very well determined). (See Figure 16.)

When the internal dynamic is not of gradient type, the theory of bifurcation is practically unknown. Nevertheless, one may expect that the gradient-like situation keeps some validity, with the restriction that generalized catastrophes may occur even in polarized media.

The list of 'ordinary catastrophes' on our space—time $\mathscr{R}^4$ plays, I believe, a very important role in the interpretation of natural morphological phenomena, whether living or non-living. The catastrophes of internal dimension two, the so-called *umbilics*, have a directing role in the 'breaking' phenomena in hydro-dynamics (breaking of waves; breaking of jets). In biology, they govern, I believe, the morphology of engulfing phenomena like phagocytosis, neurulation, and, in reproduction, the emission and spreading of gametes.

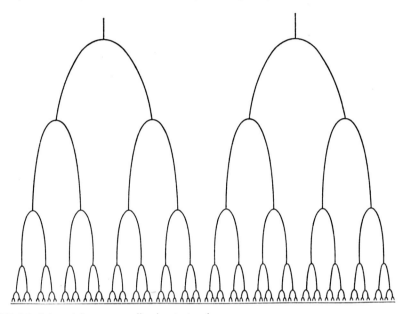

FIGURE 16 Scheme for a generalized catastrophe.

▶ *The reconstruction of the organizing center.* There is still an algebraic pheno-menon which we have to describe before introducing our biological models. This is the *structurally stable reconstruction of an organizing center.* Let us consider the cusp $4\,u^3 + 27\,v^2 = 0$ of the Riemann–Hugoniot catastrophe; $V(x) = x^4/4$, the unfolding of which is $V(x) = x^4/4 + u\,x^2/2 + v\,x$.

Inside the cusp $(4\,u^3 + 27\,v^2) < 0$, we have two stable regimes in com-petition, corresponding to two minima of the V function. Let us admit now that the local dynamic admits some component in the external variables $U = U_o$, $V = V_o$, in particular at the local attractors : to each stable regime corresponds in the (u, v) plane a vector field V, V'. Let us suppose that those vector fields are like those in Figure 17 : for the dominating regime in $V > 0$, we have V_o negative, and conversely. Put U_o positive for both V and V'. Then, if the initial position is inside the cusp, the representative point goes to 0 by a succession of oscillations with reflexion on the branches of the cusp. (Here we admit, contrary to

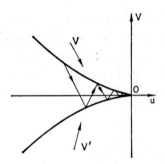

FIGURE 17
Stable reconstruction of the organizing center (Riemann–Hugoniot catastrophe).

Maxwell's convention, that each local regime persists till its complete breaking [perfect delay].) (Cf. Figure 17.)

We believe that this gives a – very simplified – model of what occurs in gametogenesis, where the organizing center of the complete somatic structure is reconstructed in the egg.

BIOLOGICAL MODELS

The static model. We admit that all possible local states of the metabolism in a living being can be parametrized by a function space, more precisely a set of potentials $V: M \to \mathscr{R}$. In this function space $L(M; \mathscr{R})$, there exists a point which represents the 'germinal state', represented by the most degenerate potential $w \epsilon L(M; \mathscr{R})$. Suppose w admits at a point $0 \epsilon M$ an isolated singularity of finite co-dimension. Let U be the universal unfolding space of this singularity. Then development of the egg may be described by a mapping $F: B^3 \times T$ of the 3-cell $B \to U$ (called the 'wave of growth') which meets transversally the catastrophe set K in U. As soon as F_t meets some components of K, new cellular differentiations appear. After some time, when maturity is reached, some part of the image $F_t(B^3)$ gets back to the organizing center w by a structurally stable process, describing (without the complication of sexuality) the formation of gametes. See Figure 18 for a global picture.

A more refined model may be defined as follows (Figure 10). The organizing center O is never realized in any point of the organism; the space U is multiplied by an 'auxiliary coordinate' y playing the role of a momentum. Put $x = |F|$ as

108

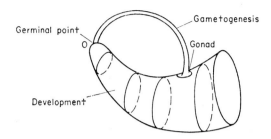

FIGURE 18

the distance to *w* in *U*. Then the wave of growth *F* describes a kind of circle of
center *O* in *Oxy* [like the trajectory of a one-dimensional oscillator in phase
space (*x, y*)]. The half circle *x* < 0 represents the haploid states (gametes),
w (*x* = 0, *y* = 1) is the fertilization of the egg. The quadrant *x* > 0, *y* > 0 represents
development; the point *x* = 1, *y* = 0, sexual maturity. The quadrant *x* > 0, *y* < 0,
gametogenesis, and the point *x* = 0, *y* = – 1, meiosis. Such a model may give
some answer to the trick question, which started first, the hen or the egg? In
fact, the 'organizing center' of the whole structure is out of the figure, and we
may consider some pathological processes, like cancer, to be a kind of approximate realization of this organizing center. (Circle c of Figure 19.)

The main defects of this model are first its imprecision; second, that it attaches
a static character to the local regimes, although they are obviously of metabolic
nature.

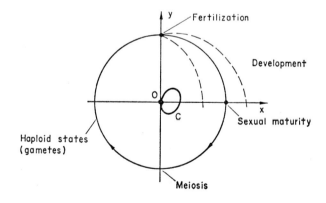

FIGURE 19

109

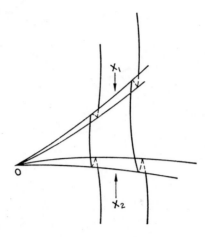

FIGURE 20
Two-dimensional representation of the figure of regulation. X_1, X_2; correcting vector fields.
0, organizing center.

▶ *The metabolic model: figure of regulation.* The global homeostatic properties
of the metabolism of any living being may be given the following geometric
interpretation : let W be an Euclidean space representing the mean states of the
organism. Suppose we submit the organism to a stimulus s; then the representative point goes to a point $s_t \epsilon W$, and then (if the stimulus is strong enough,
but not so strong as to kill the animal immediately) the metabolism gets into an
excited regime; as a result, a correcting vector field Z describing a reflex r appears
in W which brings the representative point back to the 'ground zone' G of W,
zone of rest, where the excited regime disappears in favor of a normal regime,
hence Z also vanishes. There exists a fundamental correspondence $s \rightarrow r(x)$
which associates to any stimulus a correcting reflex (or a sequence of them).
This whole structure can be generated by an unique 'organizing center' in a
multidimensional space and is called the *figure of regulation.* (See Figure 20 for
a two-dimensional scheme.)

The main postulate of our model is the following : if we describe in a convenient function space, the metabolism of the young blastula cell or the metabolism of the gametocytes — primitive germ cells — then the geometric picture
defined by the regulation of this metabolism *simulates* (in a sense which can be
made mathematically precise) the *figure of regulation of the whole organism.* As
soon as development proceeds, this figure becomes too complicated to be

110

stable. Some cells — those near the 'animal pole' of the egg — specialize in
s-states; those near the vegetative pole specialize in r-states; in advanced
animals (vertebrates) the s-cells lose any regulative power, and become nerve
cells : neurons, having lost their regulative abilities, keep track of everything
happening to them, a very important property for the future organ of memory.
The main physiological fields involve, at the adult stage, a complete sequence of
ordered chreods : for instance, an alimentary reflex implies the following sequence:
seeing a prey; capturing it; bringing it to the mouth; eating it; motor and
glandular activities of the digestive tract. All this sequence is represented, at the
blastula stage, by a preferential oscillation $s \underset{\leftarrow}{\rightarrow} r(s)$ in the metabolism, which
keeps going to some extent even after differentiation. When later in develop-
ment two tissues carrying such a truncated oscillation come into contact, a
biochemical resonance ensues, and a local regime arises through interaction
(embryological induction) : the corresponding catastrophe builds then an organ
of this chain of reflexes. Conversely, in gametogenesis, all those oscillations disappear successively with the vanishing amplitude of the metabolism; when such
an elementary oscillation vanishes, it gives rise to the condensation of a genic
material, the 'biochemical vibrations' of which restore the oscillation after
fertilization. This is the general scheme, which we cannot develop here at a
greater length [13]. The mitotic cycle itself may be described in the same way.
▶ *Spatio–temporal development.* The preceding models tried to describe only
the 'internal biochemical' variations of the local metabolism, and not the spatio–
temporal morphology they cause. In order to describe this morphology — at least
qualitatively — one makes the relatively mild assumption : to any stable local
regime, there corresponds in three-space a propagation of the corresponding
tissue described by a variational principle of the 'Lagrangian' type (each regime
having its own Lagrangian). Then the successive evolution will be described by
a kind of wave-front, and this wave-front may present singularities of the type
described by our 'elementary catastrophes' — at least initially. Quite frequently,
because of the polarization of the tissue, these geometric catastrophes are
coupled to biochemical ones, and give rise to new differentiations. For instance,
gastrulation in the amphibian egg is an ordinary catastrophe, defined by a circle
(closed curve) of 'swallow's tails' (*queue d'aronde*) separating ectoderm from
endoderm (Figure 21); the primitive streak, in birds' embryos, may also be
interpreted by the formation of a double line limited by two 'swallow's tails'
separating ectoderm from mesoderm. (Figure 22.) Of course, the early delamina-
tion of hypoblast and epiblast — in birds and mammals — has to be considered

as a generalized catastrophe, due to the insufficient polarity of the tissue when it starts (manifesting probably an increasing power of regulation for these eggs). Later on, genetic constraints do appear, the first of which is bilateral symmetry; its 'organizing center' is chord formation, and its effect is very strong on the dorsal s-directed tissues; it disappears finally on the ventrally located organs, like the heart and digestive tract. The mathematical theory of these constraints is more sophisticated : one may express it roughly by saying that the 'external (unfolding) coordinates of a chreod C_1 play the role of internal variables for the succeeding chreod C_2'. Such a rule introduces degeneracies of a more complicated type (singularities of composed mappings), and leads to a more complex morphology. This may explain why finally we get chreods which are subjected to a relatively precise metric control as in bones and eye formation.

▶ *The hydraulic model.* The following model has been inspired by an idea given by C. H. Waddington [14], the idea of the *epigenetic landscape* ; C. H. Waddington proposed to look at the development in embryology as the trajectory of a material point in a 'landscape' where valleys define the main paths of development. Here, we propose, according to our Lagrangian way of describing development, to regard it as given locally by the propagation of a wave-front in a three-dimensional domain. More precisely, if development is given by a function of the type $S(x,y,z) = t$ (where S defines some kind of 'action'), we construct a 4-dimensional landscape by putting a height coordinate $u = S(x,y,z)$, and then we consider the level varieties of the u function. We may materialize the model in three-space by shrinking the number of independent variables from three to two. We realize then a kind of potential well, the bottom of which is the germinal point $u = 0$. We then flood this well by pouring water in it. The shore of the lake so obtained describes the spatial development of the embryo ; there are three main valleys in the geography of the potential well corresponding to the three main layers : ectoderm, mesoderm, endoderm. The ectoderm valley communicates with the 'neural lake', describing the formation of the neural plate ; the endoderm valley with the sinuous meanders of the digestive tract ; and so on : each layer, having its own Lagrangian, has a specific slope in the well. Another interest of this model is to make possible a representation of (vegetative) reproduction. High above the mesoderm valley is a suspended lake symbolizing the gonad. Now suppose we have two exemplars M_1, M_2 of our potential well, M_1 being above M_2. Suppose at the bottom of the gonad lake there is a small pipe pouring above the germinal point of M_2. Then, if we fill M_1 till the level of the gonad is reached (sexual maturity), water will pour from M_1 to M_2, thus describing the

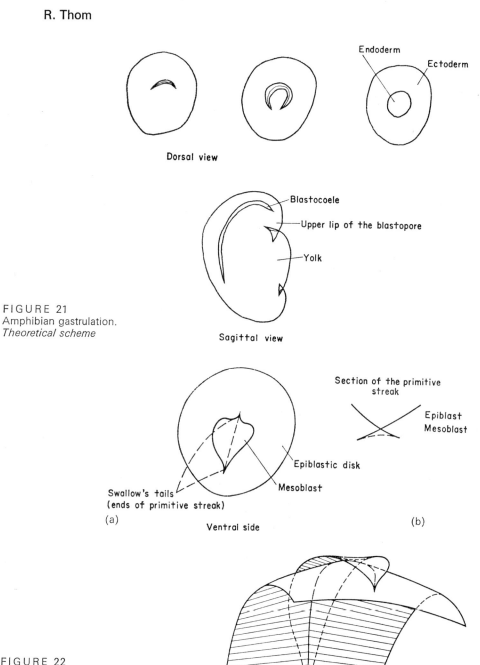

Dorsal view

FIGURE 21
Amphibian gastrulation.
Theoretical scheme

Sagittal view

Swallow's tails
(ends of primitive streak)

(a)

Ventral side

(b)

FIGURE 22
(a) Ventral side; (b) section of the
primitive streak; (c) view of the
theoretical surface associated to a
couple of swallow's tails.

(c)

113

development of a progeny child from M_1. (In fact, in the mammals, the pipe connecting M_1 to M_2 has a kind of anatomical realization in the umbilical cord.) See Plates 5 and 6. Despite the obvious shortcomings of this model (due to shrinking of dimensions, and also to the fact that singularities of wave-fronts are not singularities of level varieties of a function), it gives some reasonable intuition of the global dynamic of reproduction in the living beings. It may also to some extent represent regeneration phenomena for species like planarians, with a high regenerative capacity.

▶ *Physiological fields.* The notion of chreod has obvious applications in physiology, where it may replace the classical notion of 'physiological field'. In fact, when development is viewed as a whole, physiology is nothing but the final stage of the unfolding of embryological fields; in particular, the internal variables of nervous activities (that is, the exciting rate of neurons) can be identified with the 'external' variables of the developmental fields : as an example, the firing of motor neurons causes the extension or contraction of muscles, a specific spatial variation of muscular tissue. Among all physiological fields, those having their support in neuronic, hence mental activities, are the most interesting. A model due to C. Zeeman [15] explains how, despite the apparently discontinuous firing of neurons, continuous and differentiable models of a 'metabolic' character may be applied to describe psychic activity. It is possible to apply all the ideas of chreods, catastrophes, and so on, to the faculty of language ; the meaning of a word, that is, a *concept*, can be associated with a 'figure of regulation' quite similar to the figure of regulation of living beings ; a concept has in general a kind of animal—vegetative gradient, and a system of excited states which insures its stability. Such a viewpoint gives rise to a topological theory of signification, to a geometric interpretation of semantics. In particular, the fundamental grammatical categories (like noun and verb) can be given a topoolgical interpretation. But, for lack of space, we shall not say more on this subject. (See [13].)

CONCLUSION

Are these models amenable to experimental control? Because of their inherently qualitative character, the answer is no. Practically any morphology can be given such a dynamical interpretation, and the choice between possible models may be done, frequently, only by qualitative appreciation and a mathematical sense of elegance and economy. Here we do not deal with a scientific theory, but more precisely with a *method*. And this method does not lead to specific techniques, but, strictly speaking, to *an art of models*. What may be, in that case, the ultimate

R. Thom

motivation to build such models? They satisfy, I believe, a very fundamental epistemological need. As long as scientific laws and mathematical formulae give us a very strong control on the phenomena (as in classical electromagnetism), there is no need to worry about possible models, and we may neglect for some time our irrepressible inclination to understand by images the basic nature of the natural processes. But, as soon as we run into difficulties, or contradictions (like in elementary particle theory now), or when we feel overwhelmed by the mass of empirical data without a clear notion of the problems at hand (like in biology today), then the need arises for some conceptual guidance in order to classify the data and to find out the most significant phenomena. If scientific progress is to be achieved by other means than pure chance and lucky guess, it relies necessarily on a *qualitative understanding* of the process studied. Our dynamical schemes – with the ideas of attractors, bifurcation, catastrophes, which remind us of the old Heraclitean ideas of fight and conflict – provide us with a very powerful tool to reconstruct the dynamical origin of any morphological process. They will help us, I hope, to a better understanding of the structure of many phenomena of animate and inanimate nature, and also, I believe, of our own structure.

Notes and References

1. See for instance the theory of 'Mach's reflection' in von Neumann's complete works: Volume VI, p. 300 (Oxford: Pergamon 1963).
2. The original Mémoire of Henri Poincaré is: Sur les courbes définies par une équation différentielle. *Œuvres complètes, Vol. 1* (Gauthier-Villars: Paris 1881). A historical survey of qualitative dynamics may be found in M. M. Peixoto, *Qualitative theory of differential equations and structural stability, differential equations and dynamical systems* (Academic Press: New York 1967).
3. Some mathematical justification for this statement may be found in S. Smale's spectral theorem, quoted in II (5.15.2) Differentiable Dynamical Systems, *Bulletin of the American Mathematical Society 73*, 6 (1967) 803. This article contains a thorough and up-to-date presentation of the theory of structural stability of differential systems.
4. These considerations do not apply to the conservative Hamiltonian systems of classical mechanics; because of the invariance of Liouville's measure there are no attractors – strictly speaking – in these systems.
5. *Tossing a dice* is a familiar case of such a situation.
6. See, for instance, J. Mather, Stability of differentiable mappings, *Annals of Mathematics 87* (1968) 89.
7. We use here a well-known theorem of Marston Morse, The Calculus of Variations, in the large *Amer. Math. Soc. Colloquium Publ. Vol. 18* (New York 1934).
8. $V(x_i)$ admits the origin $x_i = 0$ as singularity of finite co-dimension, if (and only if, in the complex case) the ideal generated by the first partial derivatives $\delta V / \delta x_i$ contains a power of the maximal ideal in the algebra of $\overset{\infty}{C}$-germs of differentiable functions at 0.
9. The word chreod was introduced by

C.H.Waddington in *The strategy of the genes*, p. 32 (Allen and Unwin: London 1957).

10. C.H.Waddington, *Introduction to modern genetics* (Allen and Unwin: London 1940). Max Delbrück, Colloque CNRS: Unités biologiques douées de continuité génétique (Paris 1945).

11. D'Arcy Thompson, *On Growth and Form* (Abridged Edition, Cambridge University Press 1961).

12. C.Child, *Patterns and problems of Development* (University of Chicago Press 1941).

13. These models are described in R.Thom, *Stabilité structurelle et morphogenèse* (Benjamin : New York, to be published. Also Ediscience : Paris.)

14. For a description of the 'epigenetic landscape', see C.H.Waddington, *The strategy of the genes* Fig. 4 (Allen and Unwin: London 1957).

15. C.Zeeman, Topology of the brain, *Mathematics and computer science in biology and medicine* (Medical Research Council 1965).

The problem of biological hierarchy

H. H. Pattee
Stanford University

The central purpose of these discussions, which grew out of the meetings at Villa Serbelloni during the last three years, has been to explore the significance of theory in biology – 'theory' in the sense of general principles characteristic of the living states of matter. Since I was trained as a physicist, my first two contributions to these discussions were attempts to define the problem of reducing molecular hereditary processes to elementary physical laws. I realized that many molecular biologists had already claimed to have accomplished this. But what they have developed is a model of the cell which behaves very much like a classical machine or automaton in which the 'secret of heredity' is asserted to be found in the 'normal' chemistry of nucleic acids and enzymes. Except for the fact that these machine parts are single molecules, the implication is that parts functioning like a machine can be described as a machine, and machines are understood in terms of elementary physical laws [1].

Now while there is no doubt that the machine is an appealing analogy, and that machine language is useful for describing some aspects of living matter, this language evades the two essential mysteries of life that are evident from the physicist's point of view. In the first place, if you ask what is the 'secret' of a computing machine, no physicist would consider it any answer to tell you what everyone already knows – that the computer obeys all the laws of mechanics and electricity. If there is any secret, it is in the *unlikely constraints* which harness these laws to perform highly specific and reliable functions. Similarly, if you ask what is the secret of life, you will not impress most physicists by telling them what they already believe – that all the molecules in a cell obey all the laws of physics and chemistry. The real mystery, as in any machine, is in the origin of the highly unlikely and somewhat arbitrary constraints which harness these laws to perform specific and reliable functions. This is the problem of hierarchical control which I shall discuss in this paper.

The second mystery is how one actually describes the existing machine-like constraints in single molecules. It is one thing to believe that all molecules obey the laws of quantum mechanics on philosophical principles, and quite another thing to actually apply these laws of motion to molecules that perform such intricate, rapid, and reliable operations as do enzyme and substrate. In my first

117

paper [2], I pointed out that these molecular machines present, for the physicist, some profound problems, which go beyond the classical machine language used in most molecular biological models. In all hereditary machines, or machines which execute logical operations, the constraints must be non-integrable (non-holonomic). Such constraints can be appended easily to the classical dynamics, but they present formal and conceptual difficulties in the language of quantum mechanics. I indicated also why the allosteric enzyme must operate as a non-integrable constraint in order to execute hereditary operations. Furthermore, I emphasized that hereditary machines must rely on statistical correlations and must consequently be subject to fluctuation errors. These errors normally increase with small size and high rates of operation. Therefore, any claim that life has been reduced to physics must at least be supported by an account of the dynamics, statistics, and operating *reliability* of enzymes in terms of quantum mechanical concepts. Finally I proposed that the exceptional speed and reliability of enzymes was not simply a useful property, but the essential requirement for life. It has been a recurring conjecture among physicists that only quantum mechanical properties can distinguish this reliable behavior of life from the behavior of classical machines with which life is, we believe, incorrectly compared [3].

I had hoped that by the second Bellagio meeting I might be able to suggest how logical properties or codes of molecular dimensions in cells could be approached using quantum mechanical formalism. But this problem is closely related to the measurement problem in quantum mechanics, which requires associating the quantum concepts of a state with the idea of a classical observable. This means using both a dynamical and statistical description of the same system. As I looked deeper into this problem, I began to appreciate what many physicists and mathematicians had realized long ago, that this is not only a formal problem, but one which also involves the informal meanings of *theory* and *measurement* and their relationship [4]. Consequently, my second Bellagio paper [5] was only a restatement of the questions which living systems generate for the physicist if he tries to perform a serious reduction to the quantum laws of motion.

By the time of the third Bellagio meeting, I had made little direct progress on this reduction problem. Instead, I returned to the origin of life problem and the mystery of how the unlikely constraints of transfer enzymes or biological codes could have come into existence spontaneously. Here the question is not, *how does it work*, but rather, *how did it arise*. Without giving it much thought, I had hoped that these two questions — how an enzyme can be described as a quantum mechanical system, and how a protoenzyme might arise some five billion years

118

ago in a sterile sea — could be studied spearately. But again I began to appreciate what many biologists had realized long ago, that the more we learn about the detailed relations of biological structure to its function, the more difficult becomes the problem of how these relations between structure and function arose.

Recent approaches to molecular biology tend to produce over-emphasis on the idea that if we understand the details of cellular events, then we are more likely to understand the origin and significance of these events. This has proven to be a half-truth. We certainly need to know the mechanism, but the more details we learn, the less reasonable appears any theory of the origin of the organization which integrates these mechanisms. Function is never determined by a particular structure itself, but only by the context of the organization and the environment in which this structure is embedded. Nor is structure determined by function alone, since many different structures can perform the same function. Similarly a measuring device is never determined by its particular structure but only by its interaction with the physical system being measured. And again the structure of the measuring device is not determined only by what is to be measured, since there are many possible devices which will perform this measurement function. In very much the same sense that physical theory, however complete, does not encompass measurement, the knowledge of biological structure, however complete, does not encompass its function [6].

▶ *What is the central problem?* In approaching this long-standing structure-function question, I think it is becoming clear that we get nowhere by separating the reduction-type question, *how does it work,* from the origin-type question, *how did it arise.* These are really two ways of viewing the same basic problem. I shall call this the problem of the description of hierarchical interfaces. A hierarchy in common language is an organization of individuals with levels of authority — usually with each level subordinate to the next higher level and ruling over the next lower level. However, neither the authority nor the subordination between levels is complete. Each level has its own laws or rules which control the behavior within each level. The effect of the subordination of the lower level to the hierarchical rule is to constrain or integrate the activities of the individuals so that they function coherently.

If there is to be any theory of general biology, it must explain the origin and operation (including the reliability and persistence) of the hierarchical constraints which harness matter to perform coherent functions. This is not just the problem of why certain amino acids are strung together to catalyze a specific reaction. The problem is universal and characteristic of all living matter. It occurs

119

at every level of biological organization, from the molecule to the brain. It is the central problem of the origin of life, when aggregations of matter obeying only elementary physical laws first began to constrain individual molecules to a functional, collective behavior. It is the central problem of development where collections of cells control the growth or genetic expression of individual cells. It is the central problem of biological evolution in which groups of cells form larger and larger organizations by generating hierarchical constraints on sub-groups. It is the central problem of the brain where there appears to be an un-limited possibility for new hierarchical levels of description. These are all prob-lems of hierarchical organization. Theoretical biology must face this problem as fundamental, since hierarchical control is the essential and distinguishing characteristic of life.

▶ *The failure of fact-collecting answers.* The strongest criticism of theoretical biology is that new facts often make theory appear irrelevant. This is a valid criticism for narrow theories which try to guess *how things work* at one level, since that question is answered on any one level by simply looking in more and more detail at what is actually happening. But a theory of general biology is not simply a set of descriptions of each level. It must be a theory of the levels them-selves. Of course we must know also the detailed facts at each level, but I have very little confidence that only by collecting more facts we shall ever explain the hierarchical interfaces which created these levels in the first place.

Of course it is true also that much of the progress in both physical and bio-logical sciences has come from choosing to concentrate on one level of organiza-tion at a time. Within the bounds of each level, languages grow in precision and formality; but, paradoxically, they tend to become incompatible with the langu-ages at neighboring levels. For example, in physics, particle mechanics and thermodynamics were developed independently to describe different levels of complexity. Statistical mechanics was at first an attempt to bridge the gap be-tween descriptions of very small and very large numbers of particles. And yet, to the extent that dynamical and statistical formalisms grew more precise, the gap was not bridged, but became even more impassable, since the statistical concept requires a 'postulate of ignorance' about the dynamical variables which is in-compatible with the complete dynamical description. I believe that very similar difficulties will arise at all hierarchical interfaces in biological systems. For example, the more detailed becomes our machine description of coding or infor-mation-transfer enzymes, the more difficult, and perhaps incompatible, will be the formal quantum mechanical description of the same systems. In a similar

120

way, the more abstractly we define the logical operations of the brain which we associate with formal thought processes, the more difficult it becomes to imagine an exact physical representation of these processes. In all these cases, we manage to gain formal precision at the higher level of description only by sacrificing some of the details of the motions at the lower levels.

The root of this problem is clearly not in the detailed facts of one level or another, but in the relations between levels. Therefore in order to better understand the origin and nature of hierarchical organization, I shall not direct your attention to the detailed descriptions of each level, but rather to the division between levels which I call the hierarchical interface.

▶ *The failure of one-sided answers.* In order to see the central problem of hierarchical organization more clearly, it is helpful to look at the difficulties which arise when the hierarchic interface is viewed from only one side or the other. Viewed from the lower side of this interface, the elementary laws are regarded as the given conditions and the problem is to see how the hierarchical constraints arise to perform integrated function at the higher level. Viewed from the upper side of this interface, the hierarchical constraints are regarded as the given conditions and the problem is to see if the integrated function is consistent with the elementary laws.

Two examples of these views may be found in recent writings of Francis Crick and Michael Polanyi. Crick [7] in his book *Of Molecules and Men* defends complete reductionism in biology because, by assuming all the hierarchical structure of cells as given, he feels that all the resulting elementary motions can be explained in terms of ordinary physics and chemistry. At the other extreme is a paper by Polanyi [8] on *Life's Irreducible Structure,* in which he assumes that all the molecules obey physical laws but claims that the origin of the constraints or boundary conditions which result in hierarchical integration are irreducible to those laws. There are, of course, many advocates on both sides of the question, but these two writings, read side by side, form a remarkable example of forceful arguments ending up with opposite conclusions. But of course the reason is that Crick and Polanyi had no common ground in the first place. They began on opposite sides of a hierarchical interface.

Crick makes it clear at once that he is only talking about *how it works* not *how it got that way*. He is satisfied that ordinary physics and chemistry tell us how it works and that the theory of evolution is the obvious answer to how life got that way; but he doubts that the process of evolution is predictable. Then he goes on to say that this distinction between how it works and how it evolved

121

is what is confusing Elsasser and Polanyi in their approach to biology. From this point of view the hierarchy problem does appear more as a source of confusion rather than the central issue in biology.

Polanyi, on the other hand, points out that it is not the ordinary laws of physics and chemistry which are significant for understanding the nature of life, but rather the exceptional boundary conditions through which cells harness the laws to perform new behavior. He believes that the evolution of hierarchical control at higher and higher levels is the essential characteristic of life, but he asserts that although higher levels are dependent on the laws of lower levels, they are not reducible to them. Polanyi clearly sees the hierarchy problem as central to biology, but judges it irreducible.

To some extent it was my disagreement with molecular biologists' assertions that life was already essentially reduced to physics that led me in my previous Bellagio papers to discuss the nature of molecular hereditary processes in more elementary physical language [2, 5]. While I remain hopeful that the question, *how it works,* can eventually be described in terms of elementary physics, I do not think Crick or any other molecular biologist has given any evidence that this has been done. The experimental results of molecular biologists are impressive, but to a physicist their claims usually sound badly overstated. One misunderstanding is caused by different 'ground rules' for what is acceptable as an explanation or reduction of life to physics. Crick, along with many biochemists, tends to equate 'reduction' with duplicating certain cellular reactions in a test tube, as if 'vital principles' or 'unclear thoughts' abhor test tubes. There is also the tacit assumption that if a chemist can synthesize an active enzyme by stringing together the right sequence of amino acids, then enzyme behavior must present no 'real difficulty'. But is not the 'real difficulty' to determine the 'right sequence'? Even Crick admits that, 'We could only synthesize a good enzyme, at this stage in our knowledge, by precisely imitating what nature has produced over the course of evolution rather than designing one ourselves from first principles.' Again, with regard to replicating nucleic acids, he says, 'There is nothing, therefore, in the basic copying process, as far as we can see, which is different from our experience of physics and chemistry except, of course, that it is exceptionally well designed and rather more complicated.' But it seems to me that it is precisely this exception and only this exception which distinguishes living from nonliving matter! Therefore it is this 'exceptional design' which must be reduced to physics. 'Exceptional design' at any level is what creates the new hierarchical function.

122

To get around this problem, Crick, along with most molecular biologists, invokes a non-physical theory — the theory of evolution. They simply assert that evolution by natural selection can account for whatever 'design' and 'complication' the cell (and the biochemist) needs to make things work. There are at least three difficulties with this assumption. In the first place, there is no evidence that hereditary evolution occurs except in cells which already have the complete complement of hierarchical constraints, the DNA, the replicating and translating enzymes, and all the control systems and structures necessary to reproduce themselves. The theory of evolution presently is a theory which applies only to pre-existing, highly evolved, hierarchical organizations. Secondly, the theory of evolution does not account for the origin of new hierarchical levels from aggregations of lower level components. This could be disputed on the grounds that 'accounting for' means to explain but not to predict, and that hierarchies can be 'explained' as systems with greater 'fitness' than subsystems which do not have the integrated behavior produced by the hierarchical control. If such arguments are not circular, at least they beg the question. Living matter is distinguished from non-living matter in the final analysis, not because of its molecular structures, but because it evolves quite differently from non-living matter. Living matter exists only because it evolves in time in a course which laws of motion have not yet suggested. It is not a question of predicting details, which indeed may be impossible. But to claim reduction of life to physics we must have at least some hint that any functional hierarchical constraint can arise from the laws of a lower level. Theoretical biology should try to provide this hint.

Crick concludes by admitting that the origin of life and the nature of consciousness may be 'difficult areas'; but by looking only at *how it works*, which means accepting the hierarchical complexities of biological organizations as they presently exist, and excluding the problems of hierarchical origins, Crick tries to convince us that physics and chemistry explain everything.

Polanyi, of course, assumes that all molecules work according to natural laws, but since no one has accounted for hierarchical organizations by these laws, he claims there must be principles of organization which are not reducible to the laws of physics and chemistry. Evolution to Polanyi is '. . . a progressive intensification of the higher principles of life'. These principles, he says, depend for their expression on the laws of lower levels, but each higher level has a context of meaning which is unpredictable from the lower levels. Thus by looking only at the highly evolved boundary conditions and excluding detailed attempts to describe how even simple boundary conditions could arise, Polanyi tries to convince us

123

that the origin of hierarchical boundary conditions can never be explained by elementary laws of motion.

I use Crick's and Polanyi's papers as examples, not to make a philosophical plea for either reductionism or vitalism but, as I said, to illustrate how a hierarchical interface tends to raise two types of question. Viewed from the upper level of the hierarchy the existing constraints are generally taken for granted and the significant question seems to be, *how does it work.* The answer found from this perspective usually amounts to the discovery that the parts obey the laws of the lower level. To this extent there is reduction. On the other hand, viewed from the lower level of the hierarchy it is the laws of motion which are generally taken for granted and now the significant question seems to be, *how could the constraints arise.* The answer usually given from this perspective amounts to the conclusion that the constraints are not derivable from the laws of the lower level. To this extent reduction appears impossible.

If anyone takes care to formulate both of these questions with precision, then I think he will find that both answers are correct for their respective questions. Nevertheless, either of these answers alone has tended, for hundreds of years, to stimulate great disputes. Since the two questions arise from disjoint perspectives, the arguments are often largely polemical. Of course I do not desire or expect to avoid arguments over this discussion. All I can say is that I am not at all satisfied with either the claim that physics explains how life works or the claim that physics cannot explain how life arose.

▶ *The concept of hierarchy.* To begin I shall limit my use of the idea of hierarchy to *autonomous* hierarchies. That is, to collections of elements which are responsible for producing their own rules, as contrasted with collections which are designed by an external authority to have hierarchical behavior. I want to talk only about what might be called natural hierarchies rather than artificial or supernatural hierarchies, such as man-made machines or 'special creations' of any kind. Secondly, I shall assume that all my examples are a part of the physical world and that *all the elements obey the normal laws of physics.* This does not mean that I assume a reductionist attitude. The question of what reduction can mean will become clearer, I believe, only after we discover the necessary physical conditions for a hierarchical interface. Thirdly, I shall limit my definition of hierarchical control to those rules or *constraints* which arise within a *collection* of elements, but which affect *individual* elements of the collection. This is the normal biological case where, for example, in society a set of laws is enacted by the collective action of the group but applied to individuals of the group; or in the development of the organism, the collective interactions of neighboring cells control the growth or genetic expression of an individual cell; or in the enzyme where collective interactions of many bonds control the reaction of an individual bond.

Finally we must recognize the essential characteristic of hierarchical organization, that the collective constraints which affect the individual elements always appear to produce some *integrated function* of the collection. In other words, out of the innumerable collective interactions of subunits which constrain the motions of individual subunits, we recognize only those in which we see some coherent activity. In common language we would say that hierarchical constraints produce specific actions or are designed for some purpose.

Right here I shall stop my description rather abruptly, since in talking about 'function' I have passed over the hierarchical interface which always causes so much argument. Let me return instead to the first three conditions for a hierarchy (1) autonomy, that is, a closed physical system, (2) elements in the system which obey laws of physics, and (3) collections of elements which constrain individual elements. I want first to express these conditions in the language of mechanics so that we can see the implications of these conditions as simply as possible.

▶ *Structural hierarchies.* Descriptions of nature using the language of physics usually satisfy our first condition of autonomy by assuming a closed system. In classical mechanics the elements or particles in this system are said to have a certain number of degrees of freedom, which is just the number of variables necessary to describe or predict what is going on. Our second condition is that the particles of the system follow the laws of motion. Classically this means that, given their initial positions and velocities at a given time, and with arbitrarily high precision, the trajectories of the particles can be predicted in the future or explained in the past. But if we are restricted to classical physics there is no way in which the third condition can be given much significance, because it requires a 'collection' of particles which constrains individual particles. The implication here is that some particles join together in a more or less permanent collection, otherwise the 'collection' would only be transient and would depend crucially on the initial conditions. It was one of the serious difficulties of classical physics that there was no inherent dynamical reason why collections should ever form permanently. In quantum mechanics, however, the concept of particle is changed, and the fundamental idea of a continuous wave description of motion produces the 'stationary state' or a local time-independent 'collection' of atoms and

125

molecules. Since these local collections are constantly being perturbed, they are not really permanent, but have lifetimes which increase with the energy of the interactions which hold them together, and decrease with the thermal energy which knocks them apart. Although there are several types of bonds between atoms and molecules we will need to distinguish only two — the strong and the weak bonds. The structures held together by the strong chemical bonds will have lifetimes much longer than structures held by weak bonds.

So far our simple physical description is useful up to the level of polymers and crystals, but now we need to see how such collections can 'constrain' *individual* monomers or atoms which make up these collections. Up to this point, our description of matter is 'normal physics' at the level of atoms and molecules, but the concept of 'constraint' begins to sound as if we are introducing new rules. What is the physical meaning of a constraint? The concept of 'equation of constraint' was in fact first necessary in classical physics because of the lack of any dynamical process to explain the permanent loss of degrees of freedom of collections of matter in solid bodies. Another type of constraint is the boundary condition which limits the values of certain degrees of freedom independently of the equations of motion — for example, when a particle is confined by a box. Both solid bodies as well as walls of boxes could be considered as collections of particles which influence the motion of individual particles, and so they fulfil the second condition of our definition of hierarchy. But while we know that solids can form spontaneously from individual particles, constraints such as boxes are usually designed by experimenters with some 'higher' purpose in mind, and in this case our first condition of autonomy would not be satisfied. However, it is primarily the stationary-state solutions of the quantum mechanical equations of motion which account for permanent constraints.

From such apparently simple beginnings we can see the origin of what are often called *structural* hierarchies. The richness as well as the orderliness in all the natural patterns of collections of molecules and crystals could be described as a selective and more or less permanent loss of degrees of freedom among many elements. Many scientists and philosophers will assert on principle that such hierarchical structure is entirely reducible to quantum mechanics. As is often the case, those experts who actually study the details are seldom so easily convinced. For example, Cyril Smith [9] has pointed out that new levels of structural hierarchies usually depend on the appearance of an imperfection in the old level. But what do we mean by an 'imperfection'? Which imperfections lead to new levels of organization, and which lead to greater disorder?

▶ *Functional hierarchies.* In spite of the enormous complexity which we can find in structural hierarchies, there is still something missing. There is seldom any doubt that such structures are lifeless. What is missing is some recognizable 'function'. No matter how intricate a structure may be, permanence is not compatible with the concept of function. Function is a process in time, and for living systems the appearance of time-dependent function is the essential characteristic of hierarchical organization. To achieve function by permanently removing degrees of freedom in a collection of elements would be impossible. Instead the collection must impose *variable* constraints on the motion of individual elements. In physical language these amount to *time-dependent boundary conditions* on selected degrees of freedom. Furthermore, the time dependence is not imposed by an outside agent, but is inseparable from the dynamics of the system. Such constraints are generally called non-holonomic (non-integrable), and have an effect which is like modifying the laws of motion themselves. For example, the enzyme is not just a permanent linear string of amino acid residues, nor a permanently folded three-dimensional molecule. An enzyme is a time-dependent boundary condition for the substrate, which through the collective interaction of many degrees of freedom controls a few degrees of freedom so as to speed up the formation of a strong bond. Nor is it the essential peculiarity of the enzyme that it is a very *complicated* dynamical system. Any system with as many degrees of freedom as that is complicated dynamically. What is exceptional about the enzyme, and what creates its hierarchical significance, is the *simplicity of its collective function* which results from this detailed complexity.

To put the problem of dynamical hierarchical control in a more general way, it is easy to understand how a simple change in a single variable can result in very complicated changes in a large system of particles. This is the normal physical situation. It is not easy to explain how complicated changes in a large system of particles can repeatedly result in a simple change in a single variable. It is this latter result which we interpret as the 'integrated behavior' or the 'function' of a hierarchical organization. Thus, we find *structural* hierarchies in all nature, both living and lifeless, but *functional* hierarchies we see as the essential characteristic of life, from the enzyme molecule to the brain and its creations.

However our recognition of *function* as having to do with a simple result produced by a complicated dynamical process is not useful unless we can give some physical meaning to the idea of simplicity. The problem is that the concept of simplification is not usually associated with the physical world, but rather with the observer's symbolic representations of this world. The world is the way it is.

127

Only an observer can simplify it. In fact it is the assumption that the elementary motions are complete and deterministic that make the generation of hierarchical rules appear so difficult. The hierarchical rule is superimposed upon a lawful system which is already completely deterministic. How can this be done without contradiction?

As far as I can see, this has never been done in physics without introducing what amounts to a measuring device or an observer. Unfortunately, since measuring devices and observers are usually associated with the brain, this does not resolve the contradiction, but only substitutes a human language hierarchy, which is a harder problem than the one we are asking. I want to think of the most elementary configurations of molecules in which we recognize some simple objective function. So again the question arises : How can a lawful system of atoms which is maximally deterministic superimpose an additional functional rule or constraint upon its detailed motions?

And again, the only answer must be that *the concept of functional constraint implies an alternative way of representing the detailed motions.* But in a closed physical system there is no observer to represent the system in a different way. Therefore we are left with the idea that if we can recognize a simple hierarchical function in an isolated dynamical system, then we should also be able to recognize an internal *representation* or *record* of the system's own dynamics. Autonomous hierarchical function implies some form of self-representation. In other words, we may partially resolve the appearance of hierarchical order on an already completely ordered set of elements by saying that hierarchical rules do not apply to the elementary motions themselves but to a record of these motions. Before we look at some examples of simple molecular collections which may exhibit internal records, let us see under what conditions our own hierarchic representations of physical systems arise.

▶ *Descriptive hierarchies.* The hierarchical levels of our languages contain some of the deepest mysteries of logic as well as epistemology, but I believe they also contain a clue to the physical problem of the hierarchical interface. We have already mentioned the crucial interface between the strictly causal language of dynamics and the probabilitistic language of statistical mechanics which has produced much distinguished controversy. I shall try to avoid the intricacies of the general arguments by using a simple example as an illustration.

When we speak of the elementary laws of mechanics we mean the laws that describe as precisely as possible how each degree of freedom changes in time, given the initial conditions and boundary conditions. These equations of motion

128

are universal and apply to all detailed motions which take place in the system. In one sense, therefore, all additional information about the system is either redundant or contradictory. But if we are trying to describe, say, 10^{23} molecules in a box, it is obvious that measuring or following each degree of freedom is impossible. However, as *outside observers* we have learned to recognize and define collective properties of molecules, such as temperature and pressure, which allow simple and useful measurements on the gas in the box. It is significant that these properties were measured long before their 'molecular basis' was known, just as many hierarchical biological functions were accurately described before a 'molecular basis' was discovered. In physics it was the later discovery of the molecular dynamics which began the controversial attempts to reduce thermodynamical description to mechanical description by rigorous mathematical arguments. Perhaps these attempts can be characterized as very nearly successful – but not quite. This result is not trivial, since 'not quite proved' in mathematics is like 'not quite pregnant' in biology.

We may look at the problem as arising from the inability of the formal mathematics to predict what collective properties of complicated systems will produce simple, significant effects in the physical world of the observer. In other words, while there is no question that the detailed equations of dynamics can be used to calculate previously well-defined averages or collective properties, there is no way to predict from only the dynamical laws of the system which definitions of collective properties are significant in terms of what we actually can measure. Thus in one sense we can derive the pressure in terms of a suitable average of dynamical variables, if we are given a precise definition of pressure; but this definition of pressure is not determined by the equations of dynamics. The concept of pressure appears useful only when the dynamical system is embedded in a particular type of observational environment.

More generally we may say that a physical system which appears complete and deterministic with the most detailed symbolic representation can appear incomplete and probabilistic only with a new representation which relinquishes some of the detail. The new representation must therefore come about through the *combination* or *classification* of the degrees of freedom at the most detailed level so as to result in fewer variables at the new level. Formal reductionism fails simply because the number of possible combinations or classifications is generally immensely larger than the number of degrees of freedom. What must always be added to define a new representation is the rule of combination or classification which tells us how to simplify the details. In statistical mechanics this rule is

129

usually a hypothesis of randomness or ergodicity, but the ultimate justification for any such rule is that it results in a more useful description of the system in the observational environment in which the system is embedded.

What can it mean, then, for a collection of particles to form an *internal* simplification or *self*-representation? What is the meaning of an 'observational environment' for a system which is closed? Clearly in an autonomous hierarchy there must be an internal separation of some degrees of freedom from other degrees of freedom which become constrained to impose collective and time-dependent boundary conditions on individual degrees of freedom. While we know such integrated systems exist in cells, and can design machines which operate in this way, we are still baffled by the spontaneous origin of this type of constraint.

It is, in fact, a characteristic difficulty of hierarchical interfaces in biological organizations that their actual operation may appear quite clear while their origin is totally mysterious. The genetic code is a good example of a crucial hierarchical interface that is clear in its operation, but mysterious in its origin. One might wonder, in fact, if there is some inherent reason why a hierarchical organization obscures its own origins. Since it is one general function of hierarchies to simplfy a complex situation, Simon [10] has suggested that if there are '. . . important systems in the world that are complex without being hierarchic, they may to a considerable extent escape our observation and understanding'. Putting it the other way around, I would also suggest that 'being hierarchic' requires that the system control its dynamics through an internal record, which has some aspects of 'self-observation'.

▶ *The lowest hierarchy.* But this is only evading the question. Let us see if we can clarify the problem of hierarchical origins by looking at collections of molecules of gradually increasing complexity, watching closely for any signs of internal *classification* or *recording* processes which are the essential conditions for a simplification of the detailed dynamics. If we can imagine such collections, then we may go on to ask if this internal simplification is inherently self-perpetuating, or if there appear to be additional conditions which must be satisfied to establish a persistent hierarchical organization of molecules.

Perhaps the simplest interesting level of complexity is crystal growth. First, consider an ideal, ionic crystal growing in solution. One might try to apply our hierarchical conditions by saying that the crystal surface, with its alternating positive and negative sites, 'classifies' the incoming ions, and by permanently binding each ion to a site with the opposite charge forms a 'record' of the classification interaction. Now while this may be grammatically correct, it is really only a redundant statement. There is no real distinction here between the physical interaction of the ion and the binding site and what we have called the 'classification' and 'record' of this interaction. They are all the same thing. Furthermore, each ion's interaction is local and direct and does not involve the dynamics of any large collection of ions or any delay. Therefore, although we may call this ideal crystal an example of hierarchical structure, I would not say that it exhibits hierarchical control over its dynamics.

Let us go on, then, to a more realistic level. Consider crystal growth which is produced by an imperfection, such as a screw dislocation. This is a *statistical* process which requires more than one atom or molecule to be in metastable positions. In time these atoms would shift to stable positions if there were no further growth. But this screw-dislocation structure increases the rate of growth by many orders of magnitude, all the time maintaining its special structure even though the original collection which first introduced the dislocation has been buried deep within the crystal. In this example, I believe a much stronger case can be made in favor of calling this a kind of hierarchical control. First, the constraint which controls the growth dynamics is not simply the direct interaction between local atoms, but involves the *collection* of atoms which makes up the dislocation. Second, this collection is not the original dislocation, but a *record* of a dislocation which is propagated over time intervals which are very long compared to the rate of addition of the individual atoms. However it is difficult to distinguish a classification process in this example since all the atoms are identical.

As a third, more complicated example, then, imagine a protoenzyme made up of only two types of monomers in a linear chain. Suppose this particular sequence of monomers folds up into a catalyst which speeds up the polymerization of only one type of monomer. For this specific catalytic reaction to occur we must express the fact that the folded polymer can distinguish one type of monomer from the other, and on the basis of this distinction alter the dynamics of each correct type of monomer so that it reacts much faster. Or in other words, we may say that this sequence of monomers *classifies* its elements and *records* this classification by forming a single, permanent bond between monomers. Now is there anything wrong with calling this process a form of hierarchical control?

In so far as the polymer sequences are no longer determined directly by the dynamical laws of the individual monomers (including their inherent reactivities), but by the constraints of a special polymer which speeds up the formation of a

131

particular sequence, this might be called hierarchical dynamics. But now I think we have some problems of autonomy. First, this specific catalyst was invented by me, and although we know such specific catalysts do exist as enzymes, my invention simply evades the origin problem, as well as the physical problem of how such specific catalysts work. However, I have in mind a problem which is much more important. I think this example misses the essence of hierarchical *control*. We may indeed have in the catalyzed homopolymer a kind of simple record of a rather complex dynamical interaction, but the record has no further effect.

The trouble is that in the context of autonomous hierarchies, what constitutes a 'record' must be indicated within the closed system itself and not by what I, as an outside observer, recognize as a 'record'. Obviously to generate autonomous hierarchical control the record must be *read out* inside the system. The time-independent constraints formed by the permanent strong bonds must in turn constrain the remaining degrees of freedom in some significant way. This was the case in the previous example of screw-dislocation crystal growth, where the dislocation structure was both a record of a past collective imperfection and a catalyst for the future binding of individual atoms. Cyril Smith sees this process as requiring a new description somewhere in between the detailed dynamics of atoms and the simple, stationary averages of thermodynamics. He sees all complex structure as both a record and a framework : '. . . the advancing interface leaves behind a pattern of structural perfection or imperfection which is both a record of historical events and a framework within which future ones must occur'.

Returning to the copolymer system, we see that it may indeed fulfil the function of a record of past events, but the homopolymer record which was catalyzed does not act as a framework for future events. To provide autonomous hierarchical control, the catalyzed product of one copolymer must lead to the catalysis of other specific reactions. Furthermore, if the record is not to be lost, each catalyzed sequence must in turn catalyze another, and so on indefinitely. Now clearly such a sequential process can be divergent or convergent depending on the rules of specificity for the catalyses. Even if we assume that there is no error in these rules, a divergent record would never be recognized. One might say, in this case, that the system's self-representation is as complex as the system itself. But I think no *under*constrained system would produce such a chain of catalysts. The starting record would simply disintegrate.

Going back now to the hierarchical control in the screw-dislocation crystal growth, we may look at this example as the other extreme. Here the classification and record possibilities are trivially *over*constrained. Since there is only one distinguishable type of monomer, there can be no classification and hence no linear record. The 'record' is not distinguishable from the three-dimensional structure which is also the functional catalytic site. The same problem of *over*-constraint could, of course, occur in a copolymer system where, say, an alternating-sequence polymer acts as a tactic catalyst for the same alternating sequence. But this is the point of these examples. I want to show that even the simplest hierarchical organization requires a balance between the numbers of degrees of freedom of its elements, the number of fixed constraints, which function as a record, and the number of flexible constraints which encode or transcribe the record.

Of course from this simplest conceivable level of molecular assembly which exhibits a potential *classification–record–control* process, we should not expect to find the nature of hierarchical interfaces at all levels. Even these simple examples present unanswered questions. But in following the necessary physical steps leading from the dynamics of individual units to the collective control of individual units, I believe we can gain some insight into the spontaneous generation of hierarchical organization.

First, we see that the individual particles or units follow more or less deterministic laws of motion. These units were atoms or molecules in my examples, but we may also think of the units as cells, multicellular individuals, or population units. The 'motions' of these larger units are not as deterministic as the motion of atoms, but they have definite patterns of unit behavior. Second, there are forces between units which produce constraints on the individuals. These forces cause permanent aggregations of units which act as relatively fixed boundary conditions on the remaining individuals. By 'relatively fixed' I mean that the rate of growth or change of these aggregations is slow compared to the detailed motions of individual units. These strong forces form what we called structural hierarchies, but they are essentially passive constraints.

The third stage is crucial and, as we might expect, the most mysterious. If the fixed constraints are not too numerous, that is, if the aggregations are not too rigid, then *weak forces* become important in the internal dynamics of the aggregations and through this *collective* dynamics the aggregations can form *time-dependent boundary conditions* for the other individual units. This type of flexible or non-holonomic constraint reduces the number of possible trajectories of individual units without reducing the number of degrees of freedom. This amounts

133

to a *classification of alternatives* which leads us now to use the higher language of information or control. The specific catalyst or enzyme is the simplest example of such a dynamical constraint; but at any level of hierarchical control where there are ordinary molecules which also act as messages, or where simple physical objects are said to convey information, there must be the equivalent of such dynamical constraints which classify alternative motions by leaving a record of their collective dynamical interactions.

As we said earlier, it is in the simplicity or relevance of these records or messages that we recognize hierarchical control; but how this simplicity originates remains a mystery. In practice, when a dynamically complex system exhibits simple outputs or records of its internal motions we switch languages from the detailed dynamical description to a higher language, which relinquishes details and speaks only of the records themselves. We might think of our simplified language as an *effect* necessitated by a system that is too complicated to follow in detail, as in the case of our thermodynamic description of a gas. On the other hand, in systems which exhibit autonomous hierarchical organization, it is the internal collective simplifications which are the *cause* of the organization itself. In this sense, then, a new hierarchical level is created by a new hierarchical language. Simon [10] has come to a similar conclusion from observing a broad class of hierarchical organizations. He calls the lower level language a detailed 'state description' and the upper level language a simple 'process description'. But the fact remains that whether it is the system–observer interface in physics, the structure–function interfaces in biology, or the matter–record interface in the most primitive molecular hierarchies, these levels are presently established only at the cost of creating separate languages for each level.

▶ *Conclusion.* I have described the simplest examples I can imagine of what might be called incipient molecular hierarchies. I have used only a rough, semi-classical language, and have not even touched on the crucial question of how specific catalysis or classification processes could be described in the deeper quantum mechanical language. Nevertheless, I find the physical concreteness of these simple examples very helpful in sorting out which conditions are most essential for establishing a hierarchical interface.

What we find is that even the lowest interesting example of a hierarchical interface is beset with precisely those difficulties that we find in all hierarchical structures, namely, that each side of the interface requires a special language. The lower level language is necessary to give what we might call the legal details, but the upper level language is needed to classify what is significant. As

Polanyi [8] has so clearly pointed out, living organizations are not distinguished from inanimate matter because they follow laws of physics and chemistry, but because they follow the constraints of these internal, hierarchical languages.

It is therefore difficult for me to escape the conclusion that to understand even the simplest biological hierarchies, we will have to understand what we mean by a record or a language in terms of a lower level language, or ultimately in terms of elementary physical concepts. Physicists have worried about the inverse problem for many years. In fact a large part of what is called theoretical physics is a study of formal languages, searching for clear and consistent interpretations of experimental observations. Biologists have never paid this much attention to language, and even today most molecular biologists believe that the 'facts speak for themselves'. Hopefully, as these facts collect, biologists, too, will seek some general interpretations. All these facts tell us at present is that life is distinguished from inanimate matter by exceptional dynamical constraints or controls which have no clear physical explanation. We will not find such an explanation by inventing new words for *our* description of each level of hierarchical control. Instead, we will have to learn how collections of matter produce *their* own internal descriptions.

This study is supported by the Office of Naval Research Contract Nonr 225 (90).

Notes and References

1. All the authors on molecular biology I have read tacitly assume that the classical idea of a deterministic machine is a good physical analogy to living matter, even though living parts are more flexible than most machine parts, e.g., D. E. Wooldridge, *The Machinery of Life* (McGraw-Hill: New York 1966). No one, except Polanyi [8], points out that machines are only designed and built by man, and are therefore a biological rather than a physical analogy.

2. H. H. Pattee, The physical basis of coding and reliability in biological evolution, in (C. H. Waddington, ed.) *Towards a Theoretical Biology 1: Prolegomena* p. 67 (Edinburgh University Press 1968).

3. This idea was popularized largely by E. Schrödinger's book *What is Life?* (Cambridge University Press 1945). Also see N. Bohr, *Atomic Physics and Human Knowledge* pp. 21 and 101 (John Wiley and Sons: New York 1958); F. London, *Superfluids* (2nd ed.) vol. 1 p. 8 (Dover Publ.: New York 1961); and E. P. Wigner in *The Logic of Personal Knowledge* p. 231. (Routledge and Kegan Paul: London 1961).

4. There are so many papers on the measurement process, I shall give only two older references, N. Bohr, *Phys. Rev. 48* (1935) 696; and J. von Neumann, *Mathematical Foundations of Quantum Mechanics* Chap. V (Princeton University Press 1949); and one more recent paper, E. Wigner, *Am. J. Phys. 31* (1963) 6.

5. H. H. Pattee, Physical problems of heredity and evolution, in (C. H. Waddington, ed.) *Towards a Theoretical Biology 2: Sketches* p. 268 (Edinburgh University Press 1969).

6. This point of view, well known to classical

biologists, is brought out sharply by Paul Weiss, From cell to molecule, in (J. M. Allen, ed.) *The Molecular Control of Cellular Activity* p. 1 (McGraw-Hill: New York 1962). Epitomizing the nature of hierarchical organization, Weiss says, 'In short, the story of "molecular control of cellular activities" is bound to remain fragmentary and incomplete unless it is matched by knowledge of what makes a cell the unit that it is, namely, the "cellular control of molecular activities" '.

7. F. Crick, *Of Molecules and Men* (University of Washington Press: Seattle 1966).

8. M. Polanyi, Life's Irreducible Structure, *Science 160* (1968) 1308.

9. C. S. Smith, Matter Versus Materials: A Historical View, *Science 162* (1968) 637.

10. H. A. Simon, The Architecture of Complexity, *Proc. Amer. Philos. Soc. 106* (1962) 467. In addition to emphasizing the essential correlation between state and process languages in any functional hierarchy, Simon characterizes hierarchical organizations as 'nearly decomposable', by which he means that the state space is larger than the trajectory space. This is nearly equivalent to what I call a non-holonomic constraint.

The role of individuality in biological theory

Walter M. Elsasser
University of Maryland

Homogeneity and individuality. The purpose of this paper is to describe a theoretical method by means of which the distinction between physical theory and biological theory is analyzed in terms of the notion of individuality. Obviously individuality exists everywhere; it cannot be confined very well to any particular realm. What we claim here is that the role of individuality in biology is *essential,* in a way to be described, whereas in the physical sciences it is incidental and quite often irrelevant. We go so far as to claim that the basic difference between living and inanimate matter can largely be described in terms of the role which individuality plays in the two areas. This is avowedly a grand claim and, while I make it here without hesitation after years on this subject, it is also obvious that such a claim can be verified only by adequate observations. There are some observational investigations of a quantitative nature into biological individuality but unfortunately they are few and far between; some of them will be brought up towards the end of this paper. They are certainly not as yet of an extent to permit the construction of a theoretical edifice based cogently on observed data. Therefore the present argument rests, in part at least, on reasoning which is originally derived from physical theory. (I am a theoretical physicist by trade.) In addition, this view can be bolstered by arguments derived from a type of biological observation which is exceedingly widespread but at present almost entirely qualitative. For years now my colleagues have repeated to me something very obvious, namely, that if I have any new ideas about biology, I will be able to convince the public only by pointing out experiments which verify the ideas. In the past these endeavors were in an embryonic stage, and this made such a requirement difficult of fulfilment : there are few things more detrimental to a child than to demand that he behave like an adult. By by now, fortunately, the tie-up with experiment can be clarified. This will be dealt with, from the observational viewpoint, in the last section.

It is the essence of our method that we should consider individuality as the end point of a *spectrum* : On the far end there is simple homogeneity and interchangeability as exemplified everywhere in the chemical and physical laboratory. In the everyday work of a chemist, lack of chemical homogeneity is described as presence of 'impurities'. Note here a value judgment imposed on something

137

that is to begin with just a fact. Impurities are in general undesirable to the chemist and, similarly, when the physicist experiments with fluids or solids he usually aims at making these materials as homogeneous in their small parts as he can. But the situation is already different when we go from the laboratory to the environmental inorganic sciences. If for instance a skilled petrologist is given a piece of rock he might, after sufficient study of its microscopic texture and chemistry, come to the conclusion that it has almost certainly been found in a quite narrow region and not anywhere else in the world.

When we enter the life sciences variety becomes a dominant feature and with it individuality. The structural and dynamical inhomogeneity of the living tissue is a basic fact of observation, and at the far end of this spectrum there are phenomena unique in space—time, that is to say, biological individuals. In the past, in academic circles at least, individuality was considered as the intellectual domain of the 'humanist', and the natural scientist paid little or no attention to it. On attaching the individual to the spectrum of heterogeneity at its far end, we shall find ourselves able to treat individuality on a common broad basis with homogeneity; thus *we shall be able to construct a gradual transition between the two*. At the same time our treatment will be purely abstract, not afflicted as far as we can help it by some of the traditional, purely verbal approaches of metaphysics.

To understand better the relationship of homogeneity and individuality, and the gradual transition of one into the other, we need to recall a basic concept of scientific abstraction, that of a *class*. Instead of explaining this concept, we shall quote from a standard textbook of mathematics [1] : 'The concept of a "class of objects" is the cornerstone of logic; it is impossible to conduct logical reasoning without involving it. It is so fundamental and so familiar that there is no purpose in either attempting to "define" it in terms of less basic concepts, or listing trivial illustrations of it. Instead, we shall merely mention the abstractly synonymous notions of a "set of elements", "collection of things", or "aggregate of individuals".' We might add that in the circumstances of empirical science the elements of a class may be either concrete, material objects or else, in the other extreme, pure abstractions, for example, formal relationships. No doubt, historically speaking, the class concept arose out of biological obser-vations. Biology offers unending examples of classes in the form of species, genera, and so on. In the physical sciences we meet a more refined concept, that of strictly homogenous classes, meaning classes whose members are all exactly alike. I have proposed elsewhere [2] the term *congruence classes* for this;

138

it may be applied to a set of objects of daily use coming off an industrial assembly line, providing that trivial differences, for example, scratches, are disregarded. Such a concept is, of course, essential to the chemist and the atomic physicist. In quantum theory it becomes a mathematical necessity : the mathematics of quantum mechanics cannot even be formulated without the axiom that all electrons are rigorously alike and interchangeable, and the same for protons.

As we go to more complicated inorganic objects, say minerals, the members of a class may be slightly different from each other. In biology, these differences become quite important, and with higher organisms they are proverbial : We know that the shepherd can tell his sheep apart, and it is one of the oldest principles of scholastic philosophy (if it is not even older) that no two blades of grass are ever exactly alike. Interestingly enough, this principle was quoted by Descartes as the reason for his disbelief in the existence of atoms. If there were atoms, he says, there could only be a finite number of arrangements forming a blade of grass, and this is not compatible with the principle just mentioned, provided one admits that the total number of blades of grass is practically infinite. This is an intriguing piece of reasoning, especially when coming from a man as great as Descartes, and therefore it should be taken seriously until demonstrated wrong. It will reappear here in a modernized form. But only half a centry after Descartes Newton appeared as his successor, and ever since Newton's time physics has been the dominant science. The physicist deals with phenomena that, if they do not form congruence classes properly speaking, are very close to them. Thus, for instance, the solutions of a differential equation may be described as a class of abstractions whose members are extremely similar structurally ; they differ from each other through their particular initial or boundary conditions.

Biological phenomena are found to be singularly recalcitrant to description in terms of congruence classes. In the extreme case where we deal with *human* affairs, individuality becomes of paramount importance. The great men of history are individuals first and foremost ; any effort at retaining the essence of these phenomena in some form of classification seems clearly futile and condemned to failure from the outset. I once read the abbreviated edition of Arnold Toynbee's famous treatise on history, and I must confess that if Professor Toynbee set out to prove that there are no formal laws of history even remotely similar to the laws of physics, he has totally succeeded in convincing at least one reader ; if on the other hand he started from the conviction that there are such laws, he has wholly failed to convince me.

139

Individuality in biological theory

In spite of this, there can be no question but that there is a progression in the historical process. (We use the word 'progression' because the common term 'progress' implies a value judgment, and we are not concerned with values.) Only a dogmatically biased person could deny that there are progressions in history, sometimes upward and sometimes also downward. But it seems at the same time very clear that these progressions cannot successfully be subsumed under anything that resembles even superficially a congruence class.

But, again, simple repetitive features in human societies are of their essence; the fact that they do not appear prominently in history books does not mean that they are unimportant but that the historian is more interested in individuality than he is in repetitive processes. There are repetitive processes in agriculture, in technology, in daily labor, in social customs, and so on; if it were not for those, the conqueror whose name shines in the history books would have nothing to conquer, or to destroy and rebuild. From this viewpoint, history is an irreducible intermingling of repetitive processes which we designate generally as *mechanisms,* and of *individualities.* Note that I am using the plural of individuality; this is convenient if we are to deal with the phenomenon of individuality on a scientific basis. Since science deals with the ordering of experience, the prime tool of this ordering process is the construction of classes. An individual, strictly speaking, may be said to constitute a *one-class,* that is a class which has one and only one element. Individuality in a generic sense, if there be any, may then be defined as the class of all one-classes, and it is at once apparent that virtually nothing can be predicated about a class so general.

We can now state our concept of biological process. *We shall conceive of biological process as an inextricable mixture of mechanisms with individualities. We aim at applying this idea to all levels of biology,* from the highest to the lowest, where the lowest level is conceived of as that of biochemical dynamics.

There is clearly a colossal jump in generalizing from the example of a historical process to the point where one reaches as basic a level as biochemistry *in vivo.* Space prevents me here from retracing the not-very-simple way in which one may arrive at this extensive generalization; the ideas may be found in detail in my two books [3, 4]. A main point can be simply stated : I agree with C.H. Waddington [5] that 'observable biological phenomena . . . are ultimately explicable in terms of concepts confluent with those used in the physical sciences, remembering that the physical sciences are themselves open-ended. . . .' The key words in this sentence are *confluent* and *open-ended,* that is, no claim is made that physics is about to converge to a 'closed' system from which biology

140

would then be 'derivable'. If we adopt a view such as this we are faced with some abstract questions concerning the degree of open-endedness of physics, the existence of regularities that are confluent with, but not derivable from physics, and generally speaking the logical consistency of types of abstraction that aim at a certain measure of *autonomy* for biology relative to physics. My efforts at dealing with these abstractions are based altogether upon the assumption that quantum mechanics in the realm of ordinary chemical reactions is a complete theory, in the sense that it does not require mathematical modification. In this respect I differ from the famous physicist E. P. Wigner [6] who suggests such modifications. The search for an abstract scheme in which biology is what we may call a semi-autonomous science, which is totally compatible with the validity of quantum mechanics, ultimately led to the conception of a biological process as an inextricable mixture of mechanisms (mainly chemical) with multiple individualities. I have convinced myself that this is a rational way of formulating a type of abstraction for biological phenomena which is 'confluent' with the principles of physics. It is satisfactory from the purely logical viewpoint, that is, it is free from arbitrary *ad hoc* assumptions as well as seeming to be free from internal contradictions. At the same time it fits empirical biology very well indeed.

In the pages that follow I plan to concentrate upon two aspects of this scheme. One is purely abstract and has to do with a definition of individuality which is essentially mathematical, and with the consequences of this definition. At the end we shall briefly discuss some existing observations of biological individualities.

▶ *Effects of complexity.* We said that abstractly speaking an individual can be considered as representing a one-class, a class that has only one member. Obviously, the classes of objects or of abstractions with which the physical scientist ordinarily deals have very many members. The abstract classes of the mathematician, commonly called sets, have frequently (although not necessarily) an infinite number of members; for instance the continuum of Euclidean geometry represents an infinity of points. The operations of the calculus, differentiation and integration, are defined in terms of limits of infinite sets. There is no need to pile up examples. Now, if we admit that individuality forms an outstanding feature of biology, we can perceive a gap which, as we shall find, can be filled by looking at classes that have a *finite* membership, and hence are intermediate between infinite classes and one-classes.

Let me note in passing that this is not a call for the mathematician to abandon

141

infinite sets when he deals with problems of biological origin. What kind of mathematics is to be used should depend entirely on the specific problem. In numerous cases only technical disadvantages result from replacing infinite sets by finite ones. In other cases a transition to finite classes is of the essence, and this will certainly hold for the problems with which we are here concerned. The situation is somewhat analogous to the relationship between classical continuum physics and the atomistic approach. In most problems of traditional physics an atomistic model is totally unnecessary, but there are instances where a treatment involving atomic or molecular physics just cannot be dispensed with.

The clue to the use of finite classes in biological theory is found, somewhat surprisingly, in the tremendous *complexity* of organisms. This complexity is a fact of experience. As a first step in dealing with it we shall introduce the following abstract argument : If an object is so complex that it has a great deal of internal structure, we frequently find that only the main lines of the structure need be specified while there remains great variability in detail. The larger the object, the more ways there are of realizing its detailed structure. Even a very primitive example will impress upon us this feature. Consider the permutations of a number, n, of objects laid out in a row and individually labeled so that they are all distinguishable from each other. The number of ways in which this can be done is equal to

$$n! = 1.2.3.........(n-1).n.$$

If n is reasonably large, Stirling's formula gives, in a sufficient approximation, $n! = n^n$, and this becomes tremendously large as n increases. Thus if we have 100 numbered objects which we lay out in a straight array, we may do this in $100^{100} = 10^{200}$ different ways. Similar statements can be made about the number of ways in which any fairly large number of objects may be arranged. To take another, slightly more biological example, suppose an experimenter succeeds in measuring, say, 100 different properties, morphological, physical, chemical, etc., of some species of organism. If we assume for simplicity that his measuring instruments in all cases have 40 subdivisions, then the number of possible outcomes of this set of measurements is $40^{100} = 10^{160}$. Of course, there will be restrictions ; not all possible combinations of measurable results will appear equally often, many will appear very rarely or not at all ; nevertheless it may be shown readily that, unless the restrictions are extravagantly severe, the number of possible combinations is still tremendously large compared to the sort of numbers we are accustomed to manipulate in any other scientific pursuit.

The important point to realize next is that the actual number of members of any one class of *physical* objects is limited. This is clear as a practical matter, but we can establish it as a matter of principle if we wish, by referring to the results of cosmology: the existence of a finite universe in which astronomers estimate the sum total of atomic nuclei to be of the order of 10^{80}. Furthermore, they estimate that the lifetime of the universe is no more than 10^{18} seconds. Hence the number of distinguishable *events* that can occur in a finite universe (obtained by multiplying the number of objects, nuclei say, into the number of time intervals, seconds say) is limited correspondingly. I have given such arguments in more detail in my books and shall only indicate the basic conclusion here: When we consider systems of increasing complexity, invariably we soon reach a point where the number of internal configurations in which the system *may* exist will vastly exceed the number of *actual* samples of any one given class that can possibly be collected in our universe. If the discrepancy between the number of possible states and the number of samples is large enough, we may next assert that two members of a class will practically never be in exactly the same internal state.

The preceding argument lets us conceive of a purely formal definition of individuality, or at least of one type of it which we may call 'configurational' individuality. We say that such individuality exists whenever the number of possible configurations of the systems forming a class is vastly larger than the number of members of the particular class that could actually exist in the real universe. For systems containing many molecules this relationship is always found true in practice. It is not too difficult to conceive of forms of individuality which are dynamical rather than purely geometrical: Thus if we have a generalized automaton with a very large number of different feedback loops, then it is easy to conceive of the possible number of arrangements (or perhaps only of flow patterns in these systems) such that the number of possible patterns [6a] is again vastly in excess of the number of specimens that can possibly be available in the world. This last argument is not meant as an 'explanation' of anything biological; instead it serves to show that individuality in our abstract sense can, under suitable circumstances, be accompanied by stability with respect to time.

If one tries to support these numerical arguments by direct observational evidence, it soon becomes clear that the complexity of organisms at *all* levels, from the biochemical one all the way up to, say, the organization of mammals or beyond, is so overwhelming that one cannot take it for granted that the appearance of individualities may be ignored in any attempt to develop theoretical biology. The interplay of individualities with the determinate laws of physics

143

will be at the center of our considerations; it will sometimes be referred to as radical *inhomogeneity* of objects or of classes, as the case may be.

It is easy to understand, however, that individuality in this purely abstract sense has no *prima facie* relation to biology: thus if we determine mathematically the number of quantum states of which some crystal is capable, we find that for a piece of measurable size the number of quantum states is vastly in excess of the number of pieces of this material that we could possibly find in our universe. The main difference between this case and that of organisms lies in the fact already indicated, that the organism has a tremendous amount of internal structure at all levels. It may be shown that in inorganic systems which, as a rule, are reasonably homogeneous internally, the behavior of the system and the prediction of its future behavior depend only on *averages* that are not critically sensitive to which specific state the system is in. On the other hand, in organisms the structure and dynamics are so complex that the difference between detailed states must at some point be taken into account if we follow the dynamics of the system for more than a very short interval of time. This state of affairs creates a basic difference between the methods of the physicist and chemist (working *in vitro*) and the method on which the biologist (when working *in vivo*) must depend. As indicated, the physicists's methods of quantitative prediction are usually based upon the requirement that a system be sufficiently homogeneous in its small parts, so that statistical irregularities in these small parts *can be averaged out*. In terms of the class concept this means that the members of a class are sufficiently similar to each other, so that on the basis of the laws of atomic (molecular) physics one can construct an abstract model which represents adequately all members of the class. Such differences between the individual members of the class as must always exist according to the arguments above can be averaged out in the theoretical model, which can be used for prediction of the class. After averaging, the models represent congruence classes which are sometimes referred to as homogeneous classes.

Next, consider abstractly classes of organisms; they belong to the category of *inhomogeneous classes*. The essential property of these classes is a negative one here from the standpoint of physical theory: For such inhomogeneous classes one cannot construct a model *based on purely physical properties* (invariably derived from the study of congruence classes) which would predict adequately what happens to the members of the inhomogeneous class. We do not claim that anything goes wrong with physics when we find in Nature an organic tissue consisting of a vast variety of different kinds of molecules; what we say is that *many statements of the kind we can make for congruence classes become indeterminate for sufficiently inhomogeneous classes because some of the processes of statistical averaging required no longer yield meaningful results.* Clearly, there are many properties of any class of organisms for which averaging will be successful and about which, correspondingly, definite statements can be made ; these properties correspond to the *homogeneous or mechanistic components* of the organism. The term should not, of course, suggest that these properties are only mechanical in the strict sense, in fact the most important varieties of such properties are the chemical ones. The essential feature of the mechanistic components is that they do represent variables which result from averages in such a way that these averages are well defined and have limited scatter. Such properties can in general be duplicated by the behavior of congruence classes *in vitro*; there is also some interval of time for which predictions made on the basis of a congruence-class model remain applicable to the organism.

If we now base ourselves upon the idea that *the essential distinguishing feature of organisms* is their extraordinary complexity and inhomogeneity, we are led to study a form of dynamics in which very numerous occurrences of individuality (in the formal sense just defined) are inextricably coupled into the mechanics of the system's homogeneous components. We assume throughout that this holds at *all* levels of the organism's functioning, from the 'lowest', biochemical one to as 'high' a level as one wants to go. Just stating such an idea may lead at once to a question in the reader's mind that is exemplified as follows : If this is true, what then is the difference between a live cat and a dead cat into which the live one may turn within a very short span of time? Considerations of this sort lead one soon to recognize that, if the basic idea is correct, *the organism must be a system that is endlessly engaged in producing, regenerating, or increasing inhomogeneity, and thereby phenomena of individuality, at all levels of its functioning.* When the cat dies this production of novel inhomogeneity comes to a stop. This type of conclusion should become more and more accessible to experimental inquiry, especially on the biochemical level, as time goes on. Let us here just indicate a recurrent experience in the history of physiology and biochemistry, namely, that anything which goes on *in vivo* is always found to be far more complicated than earlier investigators had thought. Thus after tremendous labors one finds, say, propagation mechanisms of the nerve or else metabolic pathways established and then expounded in the textbooks — but a generation later the workers in the field discover that these models were grossly

145

oversimplified and that biological reality is vastly more complex. It seems so frequent an experience in biology that one is led, purely empirically, to suspect a basic problem. A recent, most impressive case is the view that all the 'information' of the organism is stored in the residue sequence of DNA, an idea designated as the 'central dogma' in the terminology of molecular biologists. Just recently, Commoner [7] has collected the evidence at present available to show how complicated the process of DNA reproduction really is, indicating that there is far more to it than replication by template. Instead, as Commoner has insisted for some years now, there are complicated enzymatic feedback cycles, which one is hard put to fit into a purely mechanistic scheme of heredity.

In the remainder of this paper the role of individuality as defined above will be discussed or, rather, the discussion will be confined to a few highlights, given such a complicated subject. We shall mention here only in passing a second general principle of which we have spoken in detail elsewhere. This concerns the limitations of measurement derived from the uncertainty relations of quantum mechanics. Niels Bohr has applied this, under the name of 'generalized complementarity', to systems of extreme complexity such as organisms. When applied to one specific organism these limitations preclude our gaining a truly detailed knowledge, by means of direct measurements, of the system's atomic and molecular variables. In practice this means that we can deal with the effects of individuality on the molecular level only in a *statistical* manner; we cannot in general specifically localize the features of individuality on the molecular level, even though we can assert on a statistical basis that they must exist. This gives to the organism some aspects of 'wholeness' (according to Bohr) and to its dynamics a degree of autonomy on which Bohr dwells particularly. Niels Bohr's writings in this field have been critically analyzed and most interesting lengthy quotations given in a recent monograph (in German [8]).

▶ *Finite universes*. The concept of individuality is natural even to the untrained mind; also, the concept of infinity appears quite early in the history of human thought. However, as we all know, a mind not scientifically trained does not clearly discriminate between very large numbers and infinity. The mental process by which one substitutes a larger, say infinite, manifold for a smaller, usually finite, one is described in mathematical parlance as *embedding*. Embedding consists in subsuming a set of abstractions of a limited size under a set of larger size so that the first set is now considered a subset of the second. This is an abstract operation so common that it is often performed quite unconsciously; the student of the sciences does, equally unconsciously, arrive at a thorough training in its use. To take an example : the continuum of Euclidean space, implying an infinite point–set, is a pure abstraction ; in any actual experience we can discern only a finite number of entities, which to our measurements appear as of finite extension ; we embed these finite bodies mentally into the mathematical Euclidean continuum.

To manipulate the concept of individuality, one requires a qualitatively different abstract process, which however may be described also as a type of embedding : the everyday notion of individuality can be dealt with by considering a given individual as one choice from a vast number of possible configurations from all of which it differs somewhat. But it appears readily that this very large set of potential configurations cannot be replaced by a corresponding infinite set, otherwise the abstract notion of individuality would cease to have any simple meaning. This point is significant enough to be illustrated by a somewhat crude example. Let us consider a historical space–time phenomenon whose individuality is beyond question. Call it N (for Napoleon, say). If it were meaningful to construct models of possible worlds in infinite numbers, each representing a finite region of space–time, then some of these, say N', N'', N''', . . ., will form what the mathematician calls a dense neighborhood around N. This means that the various models differ from N but, since there is an infinite number of such models, some of them can be found to differ from N only a little, as little as one desires. This is obviously not a realistic procedure.

If we now choose to call such a continuum an abstract *embeddor set* into which we embed the objects of our experience, then we may similarly speak of a quite different but equally useful abstract embeddor set constructed to represent individuality. This new embeddor set will consist of the finite sequence of all the conceivable distinguishable configurations of the object at hand. Such a set, although finite, is so tremendously large that the specific configuration of the object, any specific configuration in fact, is a uniquely rare event. As is shown in the branch of mathematics known as combinatorics, the property of being an exceedingly rare event will always be true when we deal with objects sufficiently complex that they are capable of a high degree of internal variability. From these arguments we see, furthermore, that the property of individuality can be brought out only if one does not let the embeddor set become continuous. (In the language of statistical mechanics, a division of phase space into discrete cells must be maintained.) As a rule a continuous embeddor set would destroy the numerical relationships which express rareness and hence define abstract individuality.

This refusal to let embeddor sets go over into a continuum or otherwise become infinite is a basic step in our abstract scheme. We do not claim this restriction to be required *a priori* or in any other dogmatic way that would be pleasing to metaphysicians. Instead, we claim only that we are indulging here in a *discourse on the method*. Our method consists in not making our embeddor sets infinite, and especially in not using continua as embeddor sets. Practical experience shows that, in all cases where systems or classes are sufficiently homogeneous, there is indeed no serious need to make a distinction between finite and infinite (or continuous) embeddors. For practical reasons, in order to apply the rigorous methods of mathematical analysis, it is often convenient to make the transition to continuous embeddor sets; but, if we do this, we lose the concept of individuality understood as an exceedingly rare configuration or event among an exceedingly large number of equivalent but purely abstract ones. Abstract constructions within which only finite sets (finite classes) are admissible will be designated as *finite universes of discourse.*

Traditionally, in the more abstract parts of physical science, we find two main types of ordering of elements within classes. In the case of determinism, the behavior of each element is given by the solution of a differential equation with respect to time. The members of such a class will be similar, apart from variations that correspond to a difference in initial conditions. This is pre-eminently the case of congruence classes. A second type of behavior commonly found in the classes of physics is *randomness.* We shall not enter into learned disquisitions about randomness here; suffice it to say that whether an observed irregularity is or is not random can only be ascertained by elaborate tests. The rigorous mathematical definition of randomness requires infinite sets; randomness can only be realised approximately in any finite universe of discourse.

At this point we replace by finite universes the infinite structures that are normally used by the physicist as embeddor sets. This does not force us to do away totally with the idea of randomness. It merely makes the use of randomness a pragmatic affair. We are safe to consider the members of a class as behaving randomly whenever the finite class can successfully be embedded into a corresponding infinite class in which randomness is mathematically defined. The emphasis here is on 'successful', and since we aim at being pragmatic we do not enter into an extensive discussion of this term. (Such a discussion could only be carried through with respect to suitable, well-defined special cases.) We propose now to introduce a third category, which we designate as *inhomogeneity,* in the ordering relationships of classes. By inhomogeneity we describe a condition that can be realized only on confining oneself to finite universes of discourse. (It may perhaps be realized in other ways on deploying a panoply of extravagant mathematics, but this would be rather too difficult to be interesting.) By means of such finite universes we are able to describe objects capable of innumerable internal configurations, only slightly different one from the other. They can be sufficiently dissimilar, however, for essential features to get lost when we limit our considerations to class averages only. (The term inhomogeneous class or inhomogeneous material has been used also in this specific sense in my previous writings.) We claim that exactly this new category of description, by means of inhomogeneous classes, is required in theoretical biology whenever we go beyond the approximation of mechanisms (meaning, by this, processes exhaustively characterized by averages which define congruence classes).

It will be useful to give at this place some extremely simple examples of finite universes of discourse. These are not in any sense meant to represent living systems : It is clearly an empirical feature of living systems that they have a high degree of structural and dynamical complexity ; in our present view this is a prerequisite for individuality without which, as we claim here, life does not exist. It becomes impossible therefore, within this scheme, to study life by means of really simple models. If one applied these nevertheless, and so sufficiently reduced the features that arise from individuality, one would be left only with the study of the numerous mechanisms that subserve life. For the moment we aim at no more than to elucidate the particular modes of abstract thought which differ in some points sharply from the habitual ones of the physical scientist, and are appropriate to the use of finite universes of discourse.

Consider the sequence of six digits, alternately one and zero, shown in Figure 1a. It is almost impossible for the physical scientist to look at this finite series without performing mentally an act of embedding, usually into a random sequence of ones and zeros of infinite length. A sequence of this type might be thought of as produced by throwing a coin an unlimited number of times. From this viewpoint the probability of the six-digit sequence, Figure 1a, is $(1/2)^6 =$ $1/64$. Concretely, this means that if one picks at random a digit out of the infinite sequence, and does so a very large number of times, then in one case out of 64 this digit together with its five followers will form the sequence of Figure 1a. Now imagine that we do not permit ourselves any kind of embedding into an infinite universe ; then no statement about probabilities can be made, as probabilities can be defined only in terms of such infinite sets. Let us try now to embed Figure 1a into a *finite* universe of discourse. This evidently can be achieved in a large variety of ways. One simple way is to use as embeddor set the totality of the 64 possible sets of six digits which contain ones and zeros in any proportion. In addition we specify that one, and only one, of these 64 sets will appear in the 'real' universe. Under these conditions we cannot account for the regularity exhibited by the pattern of Figure 1a; we may say that we have 'primary' postulates, which determine that we have a pattern (any one pattern of six digits), and a 'secondary' result, empirical so to speak, which indicates a higher degree of regularity than the primary postulates would have led us to expect.

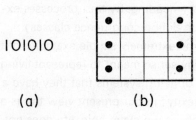

101010

(a)   (b)

FIGURE 1

A similar example, just a little more elaborate, is shown in Figure 1b. We start with twelve boxes in a 3×4 array, half of them filled, half empty. We may take this to be the result of a random selection, either from an infinite set of such arrays, in which the distribution of full and empty boxes within any one array may be assumed random, or, alternately, the array may be cut out from an infinite rectangular grid that covers the whole plane. In either case we have succeeded in embedding the given array into an infinite statistical set. Furthermore in the first of these two cases of embedding we may wish to specify that the number of full and empty boxes be equal. A simple calculation shows then that the number of different ways in which the array of 12 boxes can be half filled is equal to $(12!)/(6!)^2 = 924$. If we specify again that out of these 924 possibilities only one is realized, we cannot predict which one it will be. Then Figure 1b shows that the one which has appeared in the 'real' universe has a remarkable degree of symmetry : the filled and empty boxes form stripes, and the appearance of these stripes is neither forbidden nor implied by the primary postulates. This feature may be considered 'secondary' again in the sense of being discovered only by an empirical inspection of the given pattern.

These examples are very primitive and leave a great deal to be desired. The 'reality' mentioned before is, of course, purely symbolic, as symbolic as everything else in this abstract scheme. The prime purpose of these examples is to provide the introduction to a certain type of abstract thought : A little reflection will indicate that virtually all scientific generalizations may be expressed as a form of embedding ; our limited experiences can be given general validity only by placing them into a suitable embeddor set. The embeddor set is chosen so as to fulfil certain general rules. If these rules have a counterpart in the observations we designate them as 'laws of nature'. Note that the universality of the laws of nature resides in the abstract character of the embeddor set, not in any given data by themselves.

In traditional logic the procedure which here, following the mathematician, we have called embedding, is known as *inductive generalization*. It is safe to say that if the methods of inductive inference are applied in a systematic and quantitative manner, in the style customary in physical science, induction presents itself almost invariably in the form of embedding. The freedom in the choice of the embeddor set, which is intrinsic to this method, corresponds to the freedom and arbitrariness that all logicians know to be intrinsic to induction. The use of embeddor sets does perhaps exhibit this arbitrariness in a clearer manner than is usually met with in discussing induction.

There is one regrettable flaw in the examples of Figure 1, but unfortunately I have been unable so far to find better examples of simple, finite universes of discourse. In the models given it is easy, too easy in fact, to find infinite embeddor sets with respect to which the configurations of Figure 1 appear as events of comparatively low probability, in other words as comparatively rare events. This would be even worse if, instead of one such event, we had a set of them. In our primitive examples we have been able to avoid the idea of rare events by formulating postulates for finite universes in such a way that no probability can be assigned to an event observed. We are well aware, of course, that this crude argument, although sound in principle, will need refinement. If presently we apply related arguments to the world of organisms, it is essential that we do not choose forms of description in which classes of organisms appear as classes of relatively rare events. To admit that some phenomena of life could be described by classes of events that are rare events relative to a framework of physics would clearly be tantamount to saying that physics is false in the realm of organisms, and hence that physics must be modified. We have already mentioned that we wish to exclude such hypotheses from our arguments. The view that physics need not be modified and that biology can be thought of as extending physics by way of the latter's 'open-endedness' is, of course, a working hypothesis, nothing more. The validity of this hypothesis can be demonstrated only by its ultimate success; up to that point the idea rests on the confidence of the investigator, as do all working hypotheses.

▶ *Organismic relationships.* In what precedes we were able to define individuality by virtue of the complexity of actual systems; this complexity entails the fact that the number of possible variants of a system vastly exceeds the size of the class of material objects which may be procured in our universe. If, on the other hand, we would replace this vast but finite number of possible configurations by a continuum there would then exist, in the immediate neighborhood of any configuration, other similar ones, and it would be very difficult or impossible to find an algorithm that exhibits individuality in a simple way. Our present discourse on the method aims at showing that the conventional embedding into infinite classes practised by the physical scientist must give way, in biology, to embedding into very large but finite classes. This represents a new method because physical theories have without exception (to the best of our knowledge) taken the form of an embedding into infinite abstract universes, where a continuum is counted here as an infinite universe in accordance with conventional mathematical procedure. Physical theory gives us a representation in terms of homogeneous classes, and such classes can be assumed without difficulty to have in principle an infinite number of members. It is a characteristic of homogeneous classes that if any individuality exists in them (and in a purely mathematical sense it may be defined readily so as to exist) such individuality is irrelevant: all relevant prediction of the behavior of these classes can be based on averages in which all individuality is 'smeared out' as it were.

The classes of biology on the other hand are to be defined as inhomogeneous classes, that is, as a mixture of mechanisms with individualities, where now individualities will influence the long-term behavior of the systems. On this basis one can define clearly the distinction of physics and biology: Physics is the study of phenomena that can be adequately described in terms of homogeneous classes (where in practice nothing is lost or gained by an embedding into infinite abstract universes). Much of biology can also be described in this fashion; this part of biology deals with the mechanistic components of the organism. However, to the degree in which biology is semi-autonomous we claim that the description of organisms in terms of inhomogeneous classes cannot be fully derived from *any* knowledge of homogeneous classes. The assumption that such a 'reduction' cannot be achieved permits us to formulate the aim of theoretical biology in a clearcut fashion: *It is the prime function of theoretical biology to analyze observational data arising from classes of organisms in terms of the abstract relationships between a given inhomogeneous class and corresponding homogeneous classes, the latter defined in terms of the laws of physics.* (*Postulate I.*) One is entitled of course to consider the analysis of homogeneous classes, that is, pure mechanisms as far as they occur in organisms, also as a task of theoretical biology; if so, then the preceding sentence will refer only to that part of the behavior of organisms that cannot be fully 'reduced' to physics and chemistry (that is, to description by homogeneous classes).

This notion of theoretical biology would clearly be void of content if one could not find that, in the inhomogeneous classes which represent organisms, there are regularities that do not have an exact counterpart in the corresponding homogeneous classes. The nature of these regularities will occupy us presently. In my book of 1958 I proposed the term 'biotonic' for such regularities. Since little interest in this terminology was apparent, I have since changed to the term *organismic*, often employed by biologists. It is, however, necessary to use the term here in a somewhat narrower and more precise sense than the conventional one. The biologist, von Bertalanffy [9], for instance, who speaks extensively of organismic features, uses the term to designate any observed tendency of the organism toward strengthening its own unity and toward the stabilization of its own morphological and physiological properties. But we know that any such aim will be achieved in part by purely mechanistic ends, in particular by means of suitable feedback cycles, where the pertinent processes are as a rule chemical. We propose here to use the term organismic only for those features of living systems that are rooted in their inhomogeneity and that therefore cannot be fully represented by mechanistic models. This obviously imposes severe limitations on any speculation concerning the nature of organismic relationships.

In the examples of the preceding section we defined some simple finite universes; we found then that, in addition to properties of these universes which followed from the basic postulates, there were certain regularities that could be observed in special realizations but could not be derived from the assumptions. The models are too unrealistic to be directly applicable to biology, but we can apply a similar type of reasoning here: The primary laws are the laws of physics which can be studied quantitatively only in terms of their validity in homogeneous classes. There is then a 'secondary' type of order or regularity which arises only through the (usually cumulative) effect of individualities in inhomogeneous systems and classes. Note that the existence of such an order need not violate the laws of physics. This can be so because, according to the analysis given above, individual events are such that very little can be predicted as to whether one particular event or another will occur. Violations of physics would exist, of course, if many individualities could be described as 'rare events' within an operationally verifiable universe of discourse. It is desirable, therefore, to introduce a specific postulate that excludes statistically improbable events from playing a systematic role in theoretical biology. Unfortunately, even scientists do not always realize that the probability of an event can *only* be defined with respect to a given abstract universe of discourse. (Thus the probability that Mr Jones will live another year may be very high according to the life insurance mortality tables, if this be the universe of discourse used as embeddor set; on the other hand, if Jones has cancer in an advanced stage the best embeddor set to use is one of comparable cancer cases and the probability may become quite low.)

After all these reservations have been made, we may try to enunciate a second part of our working hypothesis about the relationship of biological and physical systems : *Individualities, configurationally or dynamically understood, feed into the dynamics of inhomogeneous systems in such a way that definite organismic trends become observable. The individual occurrences cannot be identified as rare events within any homogeneous class of physics, nor can the existence of the organismic trends be either proved or disproved on the basis of results derived from homogeneous classes. (Postulate II.)*

The two postulates given above are in the nature of a heuristic scheme which outlines the kind of biological theorizing that we may expect to be successful. The verification of these postulates will constitute a huge task; only a fool would consider verification feasible within the confines of a strictly limited stretch of work or a 'crucial experiment'. This task will require not only the elaborate methods of mathematical statistics; these in turn are merely tools suited to deal with the observed pervasive inhomogeneity of living material. I have no doubt that this approach to theoretical biology will very gradually make its way; I am upheld in this belief by the realization that to all appearances this direction of thought is, at least in its general outlines, uniquely determined, provided only that one admits that quantum mechanics constitute a definitive theory for the energy range significant in organisms. One could hardly have recognized this direction *a priori* ; however, it emerges more and more clearly after one has proceeded far enough with the analysis. Let us say here that arguments formulated by the famous mathematician von Neumann (which I have discussed at length elsewhere) demonstrate clearly that no representation of classes using only embedding into infinite abstract universes can possibly lead to a semi-autonomous biology. True, von Neumann's verbal terminology is at first sight different from the one used above, but one may verify without much difficulty that his main result is equivalent to the one enunciated. We are then led to the use of finite universes of discourse as the tools of abstract description; once one has realized the value of this abstraction, things begin to fall into place, and we may well hope that an abstract foundation of biological theory can be established by proceeding from here.

We may say next a very few words about the specific nature of organismic regularities looked at abstractly. It will appear even from the summary postulates just given that organismic regularities are to the farthest remove from the precise quantitative regularities familiar to the physicist and chemist which can be expressed in terms of congruence classes. This statement needs little explanation, since organismic regularities arise only from the multiple couplings of individualities into the variegated mechanisms available in the utterly complex systems of biology. Organismic regularities are therefore best described as *trends* that result from the coupling just indicated. (The student must here be asked to liberate himself from the frequently encountered prejudice, namely, the belief that whatever cannot be accommodated in the rather inflexible scheme of congruence classes is not 'science', or is at the least not 'true' science.) One of the chief trends found in the world of organisms (though by no means necessarily the only one) is the trend toward stability of individuals, as well as classes, with respect to time. Here we find in the theory of finite classes a formal apparatus that expresses abstractly the qualitative ideas lying behind the viewpoint of the 'holists' or 'gestaltists'. There is no intent whatever in this to disregard the mechanistic means of stabilization, specifically feedback couplings; but the main question raised by the mechanistic school of thought, namely, whether the mechanistic devices are by themselves enough for stability, *cannot be settled by purely abstract arguments*; this would be possible only if we could be assured that there existed an infinite real universe from which we could draw an unlimited number of samples of any class of objects whose existence is compatible with the validity of quantum theory. Failing this availability of infinite classes, the *empirical* consideration of the case corresponding to a finite real universe and finite abstract universes of description must needs involve individualities which cannot be predicted on general grounds; thus a decision on the correctness of the point of view given here will ultimately rest on empirical observations.

155

Individuality in biological theory

A trend that cannot be formally reduced to congruence classes by the expedient of eliminating 'noise' (and individualities, by their very nature, cannot be gotten rid of) constitutes a notion rather foreign to the physical scientist. On the other hand, this is patently the situation one is accustomed to in that part of the study of 'life' which is concerned with human affairs. The clinical report of a medical man, the treatise of a historian or a sociologist, the analytical findings of a psychiatrist reveal *trends*, not quantitative regularities expressible in terms of congruence classes. This should make clear the nature of the method of which we avail ourselves here; it may be described as follows : Physical science has been so eminently successful that very much of biology consisted in *extrapolating* the procedures, methods, and types of results of physical science into biological situations. Such extrapolation has also been eminently successful in biology, as attested by the admirable results of what is so often called 'molecular biology'. Limitations of this method arise, however, owing to the finiteness of our universe and the ensuing requirement that our description be in terms of *finite* abstractions, a restriction which leads to the appearance of abstract individualities, as we have seen. We claim next that cumulative actions due to individualities are capable of producing trends of the organismic type. We will claim also, therefore, that the trends observed in, generally speaking, human affairs are *prima facie* expressions of the abstract potential for such trends that can exist in finite universes of description. This leads us to propose that one should approach biology not just by extrapolation from the side of physical science, from 'below' as it were, but by *interpolation,* coming from both 'below' and 'above'. If we assume, again hypothetically, that such trends are a distinguishing characteristic of living matter at *all* levels of its organization, then clearly we must admit also that they are found at the biochemical level. This view seems to be close enough to the ideas on the non-mechanistic character of basic biochemical processes which have been suggested by Commoner [7]. Since Commoner's work rests on a sound experimental basis, one may be hopeful that out of further analysis of this type, combining both an experimental and a theoretical approach, new insights will emerge.

At the risk of being very pedantic we should reiterate once more the importance of the basic formal difference between the abstract concept of randomness and that of inhomogeneity, as the latter is defined here. Randomness can be defined in a rigorous mathematical fashion only by means of infinite sets where the average (expectation value) of any quantity whatever is well defined. If this concept is applied to quantum theory then, as von Neumann has proved, such constructions do not have the degree of 'open-endedness' required for a semi-autonomous biology. The concept of inhomogeneity, on the other hand, as used here refers to a finite but exceedingly large set of distinguishable configurations of a given type of system; the essential novelty here lies in the fact that we assume only a subset of negligible extent to be 'real' (although this subset may be very large indeed in absolute numbers). We can then no longer form all possible averages within the 'real' set; moreover the notion of individuality can now be employed in a purely abstract form. This supplies a theory that has just the required degree of 'open-endedness' to permit of organismic trends that express the effect of multitudinous individualities coupled into mechanisms.

If this approach prevails it will also have repercussions on the views held by philosophers (or by those scientists who want to enter into philosophical arguments). Any biological philosophy must have a strong *nominalistic* component. The term nominalist arose out of arguments about the nature of concepts that have been going on among philosophers practically ever since the days of the ancient Greeks. A nominalist is one who considers a concept merely as the name for a class which has been constituted as a collection of similar objects. Opposed to this is the Platonic notion that classes are the imperfect realization of ideas which have an independent existence in a metaphysical sense. Then there is the more modern notion of a congruence class whose elements are strictly identical by virtue of laws of nature. In both the last two cases, the concepts of a class is often referred to as a *universal*. The reader will notice, however, that I have not said that I am a nominalist to the exclusion of anything else; I merely say that it is important to maintain certain nominalistic aspects of biological thinking; although I cannot enlarge on this problem here it seems useful to emphasize the idea in view of the strong mechanistic intellectual tendencies seen everywhere which, by their very nature, tend to put too much stress on the use of congruence classes, that is, philosophical universals.

▶ *Observational individuality and inhomogeneity.* In what precedes we have presented arguments of an abstract nature dealing with the definition of individuality, and also with inhomogeneity, where we defined the latter as a coupling of the effects of multiple individualities into the mechanisms (chemical or other) that are ever-present in living bodies. We found that we could arrive at a purely abstract definition of individuality by the device (itself purely abstract) of changing from a mathematically infinite universe of discourse to a finite universe of discourse. It stands to reason that the utter complexity of living systems and the resulting immense number of the possible internal configurations of each inhomogeneous class precludes in practice a precise mathematical treatment such as the physicist is accustomed to for the simpler instance of homogeneous classes. In this respect the situation in biological theory is somewhat similar to that prevailing in chemistry, where mathematical analysis is also out of the question except for some utterly simplified paradigms. On the other hand, in dealing with chemistry we barely expect major conceptual novelty, feeling confident that description in terms of randomness and congruence classes is adequate. In biology we meet a type of conceptual novelty which is represented by finite universes of description. Since, however, this description is not conducive to results of a mathematical nature that could be made applicable in practice to specific cases (at least not outside of the domain of legitimate mechanistic models) one may well ask what all this accumulation of abstraction is good for. The answer is simple : it is meant to afford protection, the protection one urgently needs against the morass of verbiage and the fogs of metaphysics which, as experience shows, do represent a serious danger to our thought unless we have assured ourselves in advance of the structural solidity of the abstract apparatus applied in the theoretical reasoning. Once we feel fairly satisfied in this respect we can begin to do justice to the utter complexity of biological phenomena on reviewing them in a suitable qualitative or at best semi-quantitative manner (the quantitative aspects being now mainly statistical) which differs sharply from the quantitative and ultimately precise numerical agreement between observation and theory that is the standard of the physicist's work. In the remaining pages we shall be limited to a very few concrete applications of the notions of individuality and inhomogeneity.

First of all I should make it clear that these ideas are far from being in any sense my own invention except perhaps for the abstract device of using finite universes of discourse introduced in the first of my two books [3]. In the year this came out (1958), and unknown to me at the time, there appeared the report of a symposium of twelve leading American biologists that had taken place in 1955 and was published under the title *Concepts of Biology* [10]. The meeting dealt with a variety of topics, some of them rather of an organizational bent, but also with some of the basic philosophy (or epistemology) of biological science. There was general agreement among these men that life is best defined in formal terms as 'ordered heterogeneity'. The reader of the present article will perceive at once the verbal similarity of this expression with the ideas propounded here. Just how deep this similarity goes can only be ascertained by extracting as much operational meaning as possible from these arguments. I

158

have given a far more detailed survey of reference [10] not so long ago [4, section 1.21], and have come to the conclusion that, as far as terms can be defined, my approach is no more nor less than a rather abstract formulation of what these scientists have been expressing. No serious discrepancy has appeared to me, either among the opinions of these men or else between these and my own. There seems to be no need for repeating my previous writings, and I refer the reader to them should he be curious about details.

We shall next say a few words about the direct observation of individuality in biological objects. Let us note that if we speak of 'configurational' individuality we will have in mind, if nothing else is said, a specific configuration of components at a given instant. This, while it unquestionably occurs all the time, is not of much interest in itself because it does not exhibit any *relationship*, and science deals with relationships. What we can often observe readily, beyond mere instantaneous individual configurations, is the *stability* with respect to time of individual features of an organism, or else that of the characteristics of a class. Here we shall only speak of the former, that is of single organisms as individuals. We may speak of dynamic individuality when there are features of the organism that have a relationship with the corresponding features at different instants of time, this relationship being simply one of near-constancy in time. This definition corresponds to the term individual as it is used in everyday non-scientific language. We should not, however, be induced by this manner of speaking to think that 'dynamical' must be synonymous with mechanistic, that is, that organismic components can be thought of as appearing only at one initial time and that the maintenance of individuality thereafter is wholly a matter of identifiable physico-chemical feedback stabilizers, and so on. As we can never cease to emphasize, such a viewpoint would impose a specific philosophical bias upon the analysis of the phenomena related to extreme complexity.

Morphological and physiological stability exists simultaneously with variations from one individual to the next of a species or subspecies; it is familiar enough from everyday life. We rarely have any difficulty in keeping the faces of two persons apart (except when they are identical twins); on the other hand, we usually have no difficulty either in recognizing the same person in two photographs taken quite a number of years apart. Similar properties involving variation among several or many individuals conjoined with stability within one given individual are extremely widespread, at least in the case of mammals. Every doctor and every naturalist is familiar with the tremendous variation of anatomical structure among members of species of mammals. This is only very rarely

159

reflected in the drawings given in standard textbooks. The literature on such variations is widely scattered and all but inaccessible to a non-specialist.

We are somewhat better off in the case of a physiological phenomenon : The book *Biochemical Individuality* by Roger Williams [11] gives an account of the extensive experimental program carried out by him and his school. Williams investigated the concentration of a number of minor biochemical compounds in the tissues of either man or some other mammals. He studied this concentration on the one hand as a function of time in the one individual, on the other as it varies among several individuals of the same species. His general result, verified in a large number of cases, is highly surprising : There is first a certain variability in the concentration of some given compound within a tissue of just one individual if observed over a sufficiently long span of time, such as days, weeks, or longer. This variation may be typically of the order of, say, a fifth to a third of the whole amount. If, on the other hand, one measures the variability of the same compounds among a group of individuals of the same species, then one finds that the variability from one individual to the next is many times larger than that found within a single individual as a function of time. Inter-individual variations by a factor of five or of ten are by no means uncommon, and even variations by as large a factor as twenty or more have been found among individuals which by all available evidence are healthy and 'normal'. The material investigated by Williams and his students is fully large enough to establish firmly the widespread existence of this kind of pattern in man and in higher mammals.

Somewhat older than Williams' data and much more extensively studied, although in a more indirect way, are the chemical variations that form the basis of immunology. There, one is dealing no doubt with compositional changes, enhanced by conformational ones, of the protein molecules that are the principal working tools of the organism. We know, of course, that each organism tries to develop antibodies against proteins of any other organism whenever the need arises, and we know also that this offers an extremely sensitive tool for recognizing degrees of kinship as well as of broader phylogenetic relationships. Although the matter is patently of great importance to us here, I do not like to exhibit to the reader my personal ignorance by trying to wax technical. I shall make free to say, however, that while specialists in immunology and related fields do mention the term individuality often enough, it has been my impression that they rarely inquire into the meaning which such ubiquitous individuality might have within the framework of a broad biological theory. This is a point where we can justifiably expect that our general theoretical ideas will prove helpful.

160

Another field where this pattern of inter-individual variability conjoined with stability of the individual has been found, is haematology. An experienced haematologist has assured me that the number of subgroups of the well-known major blood-groups which can be identified by sufficiently refined methods, is quite large. As a result (on the mathematical grounds indicated previously) the number of blood-group patterns identifiable on this refined scale vastly exceeds any conceivable number of human individuals, so that there exists individuality of blood-groups in the precise sense of the term given above. According to my specialist colleague one could thus use blood groups just as well as fingerprints for the identification of individuals except for the difficulties of technique and the correspondingly high costs.

This brings us to one more field where the combination of inter-individual variability with individual stability prevails, that is, the subject of fingerprints. This is, of course, of a type that would be classified under anatomy rather than physiology but, apart from this mainly semantic matter, the conditions of variability and individuality are quite similar to those described in the previous examples. They are also well enough known even to the layman (among whom is this writer) so that we can save the reader a further discussion of this subject. There is one distinct advantage to this field, namely that our technical knowledge of it is so very extensive. The coexistence of inter-individual variation with individual stability has been established here and known on the broadest possible basis for many years.

The phenomena just described, comprising variability concurrent with stability, to be abbreviated now for simplicity as *variostability*, were exemplified in four physiological or anatomical forms: biochemical in the specific sense of R. Williams, immunological, haematological, and in the structure of fingerprints. It would be a reckless scientist indeed who dared assert that these four instances are isolated and unique; in fact one may assume with a good degree of confidence that variostability in the sense described is a most widespread property of higher organisms. Some of our preceding general speculations indicate that variostability, or at the least phenomena that are closely similar, are a very general feature of organic life. Undoubtedly the details of the phenomena will be different at different organizational or evolutionary levels. (Insects, for instance, show rather little variance of individuals within a given species, but tremendous variability among the species, there being about 700,000 species of insects as against 4,500 of mammals [12]. However, the correlation of number of species with any evolutionary order is not regular; it could in fact hardly be more irregular than it is found to be.) We may safely expect, furthermore, that whatever is equivalent to variostability at more primitive levels of the evolutionary scale and in more elementary activities of organisms will express itself primarily in biochemical terms. We already have intimations of such behavior in Williams' results on biochemical variability in man and mammals. But one would find it hard to believe that such a remarkably distinct phenomenon should make its appearance suddenly at the higher evolutionary levels without an appropriate equivalent at the lower ones.

The question that will be foremost in the minds of those who try to advance biological theory is clearly whether variostability can be represented in terms of purely mechanistic models. The principal mechanisms known that subserve stability are on the one hand feedback (homeostasis) in appropriate forms, on the other the repeated 'copying' of information from its storage in a stable 'template', specifically DNA. Now it is undoubtedly possible to explain much of the observed physiological and anatomical stability of organisms by means of devices or processes of this kind. But it is also clear that problems pertaining to such explanations become infinitely more involved if we do not concentrate just on isolated cases of stability but, instead, combine them with pervasive inter-individual variability in the manner described. We shall leave it to the imagination of our reader to see how he fares if he tries to conceive of a set of mechanistic stabilizing devices which not only maintain over a long time features of a single individual, but also do the same for the corresponding but different features of a vast number of individuals, members of a species, each of which, as experience shows, maintains its own unique characteristics over a long span of time.

We have now arrived at a point where the promised *observational approach* toward the distinction between a purely mechanistic biology and one in which life is considered as semi-autonomous can be stated clearly. Variostability can be directly observed in a vast number of cases, and this can certainly be done in a way which allows us to combine it with an exhaustive study of the purely mechanistic stabilizing devices that apply in each case, so as to show whether or not these are sufficient by themselves for stability. In order to get the most out of such a study, the observing biologist will have to acquire some technical knowledge of statistics, or possibly team up with a mathematical statistician, but this again does not constitute a condition of insuperable difficulty. A good experimentalist is unthinkable without a substantial dose of imagination, and for that reason I shall not presume on the imaginative powers of younger experimentalists by trying to give them examples of how my own limited mind conceives of such

162

a task. The task on the whole seems clear enough. It is the more significant as the realm of variostability is intimately related to, being perhaps better described as an expression of, the 'ordered heterogeneity' which was quoted before as the concept that grew out of the observations of practising biologists [10]. I find it extremely likely that, on suitable extension guided by empirical facts, variostability will overlap, and will perhaps become largely coextensive with, the abstract notion of the radical inhomogeneity of systems and of classes upon which all our previous ideas were based. Hence the observational study of variostability (along lines that are in part already laid out in the limited existing literature) is likely to be the empiricist's most promising approach if he wants to study the semi-autonomous aspects of biology.

At this point and even through much of what has preceded it is easy to hear the voices of those who will tell us that such theoretical subtleties might be attractive to some people who have nothing better to do, but that they are of little avail to the hard-boiled practical man. We should emphasize therefore that our theoretical analysis is intimately related to one of the most venerable traditional controversies of practical medicine : This is the one between treating specific diseases on the basis of symptoms and remedies considered under the viewpoint of established typical pathology, and alternatively treating the whole man, the patient as an individual. In practice this is usually a matter of how material, selected from the endless available amount for the benefit of the student of medicine, is to be divided among its more mechanistically and its more organismically oriented aspects. If progress in the training and practice of medicine based on such theoretical results as intimated above should be even moderately significant, one would be hard put to it to deny to them practical value as measured on any conventional scale.

In conclusion, we should amplify the observational approach toward biological anatomy just discussed by drawing attention to another area where organismic phenomena appear fairly accessible to study. (I have pointed this out before, especially in one paper [13].) One arrives at the underlying idea by means of a simple piece of reasoning : If individualities, in another term inhomogeneity, is essential to organic life, then it would be somewhat surprising to find that the organism gives itself over to the appearance of individualities just by way of accident, as it were. Instead, we may suspect that living beings are built and function in such a manner *that the dynamics of the organism subserves* (among other things) *the sustained generation of new inhomogeneity*. This may be the case at every level of organization ; for instance, thinking of sexual reproduction

163

primarily as a genetic 'mixing' device is a view so well known that on mentioning it one feels almost like stating a platitude. Such phenomena should be particularly interesting at the biochemical level, because there they may contrast distinctly with the otherwise so often postulated determinate nature of biomolecular processes.

One aspect of this problem that is rather easy to exhibit concerns the relationship of the typical biochemical dynamics to 'information stability' within the organism. (Fortunately we do not require here a precise definition either of information or of its stability; the approximate conceptions that are being discussed so widely will be sufficient for present purposes.) There is one basic and inevitable feature in the design of (usually electronic) computing machines, namely, the need for the protection of information from noise. This aim is normally achieved by making the information signal very large compared to the noise signals; this will 'buffer' the information from noise. But there is absolutely no evidence that the organism operates on principles which would embody such buffering; all the evidence is to the contrary. Chemical reactions in organic tissue are iso-energetic within small margins; if the transfer of a large quantity of chemical energy is involved then two or more reactions with energies of different signs will be coupled in such a way that the *net* loss of energy in the total reaction process is small; it is not as a rule very large compared to the mean thermal energy. One can readily understand that the mere preservation of information in static form poses less stringent requirements than does the requirement of preserving information while it is being processed; that is, during transformation processes specifically, the information needs to be buffered from deleterious noise, and especially so if the two are of comparable orders of magnitude, as they are in biochemistry. We have dealt with this aspect of biochemical dynamics earlier [3] and here can mention it only.

Another distinguishing feature of organisms as compared to engineering structures is the vast variability of their biochemical components and the complexity of the mutual interactions of these components. Thus in complicated organic molecules isomerism and in particular the resonance of conjugate bonds produce a tremendous variety of molecular configurations that are nearly equivalent energetically. If we add to this the vast number of molecular configurations that can result from the great mobility of protons and from the effects of the long-range electrostatic forces, we may estimate easily that the total number of microstates of a cell with distinguishable characteristics having an effect on its later development is vastly larger than the total number of actual cell-configurations that can possibly be encountered in a finite universe. These are again the conditions characteristic of formal individuality in the sense outlined before.

These biochemical relationships exhibit a remarkable similarity to the phenomenon of variostability as just described. But in the last-named case the stability of individuals with respect to time was embedded into the variability from one individual to the next within the species; in the case now considered we are dealing again with the stability of individual features in time, but now these are embedded into the almost boundless variability in time of the biochemical structures and processes within the organism itself. The formal similarity of the two cases is apparent. Whether we want to speak here of another kind of variostability or use different terminology is of minor importance. We want to point out that this type of relationship, where 'information stability' and tremendous variability coexist in this peculiar manner, foreign to mechanistic models, lends itself outstandingly to observational and experimental studies. Here again there is a certain overlap between the work of the biochemist or biophysicist and that of the statistician. Once more I will refrain from giving good advice about particulars to those who want to embark on such experiments and who thus may help us find the way to an ultimate decision between the ideas of a purely mechanistic and those of a semi-autonomous biology.

In dealing with these phenomena, among which variostability is particularly susceptible to observational or experimental study, we are indeed in the realm of the organismic; we observe behavior patterns that do not contradict physics but that at the same time cannot be described adequately in terms of mechanisms (the description thus cannot be subsumed under congruence classes). This brings us back to our starting point. I have dealt at length in my previous writings with the logical difficulties in the construction and understanding of an organismic theory. The observational study and verification of an organismic theory cannot be done in one sweeping operation as we have shown; on the other hand by a more gradual, cumulative process organismic biology can be brought within the scope of observational investigations.

The writer is indebted to the Office of Naval Research for financial support of his work.

References

1. G. Birkhoff and S. MacLane, *A Survey of Modern Algebra* (Macmillan: New York 1953).

2. W. M. Elsasser (to be published).

3. W. M. Elsasser, *The Physical Foundation of Biology* (Pergamon Press: New York and London 1958).

4. W. M. Elsasser, *Atom and Organism* (Princeton University Press 1966).

5. C. H. Waddington (C. H. Waddington ed.), *Towards a Theoretical Biology 1: Prolegomena* (Edinburgh University Press 1968) p. 103.

6. E. P. Wigner, *The Logic of Personal Knowledge* (Routledge and Kegan Paul: London 1961) pp. 231–8.

6a. Compare S. A. Kauffman, J. Theoret. Biol. 3 (1969) 437–67.

7. B. Commoner, *Nature 220* (1968) 334–40.

8. K. M. Meyer-Abich, *Korrespondenz, Individualität und Komplementarität* (Franz Steiner Verlag: Wiesbaden 1965).

9. L. von Bertalanffy, *Problems of Life* (C. A. Watts & Co.: London 1952). Reprinted as paperback (Harper Torchbooks: New York 1960).

10. R. W. Gerard (ed.). 'Concepts of Biology' in *Behavioral Science 3* (1952), 92–215; reprinted as Publ. 560 of the National Academy–National Research Council, available through their Publications Office.

11. R. J. Williams, *Biochemical Individuality* (John Wiley and Sons: New York 1956).

12. Lord Rothschild, *A Classification of Living Animals* (John Wiley and Sons: New York 1961).

13. W. M. Elsasser, *J. Theoret. Biol. 3* (1962) 164–91.

Remarks on statistical mechanics and theoretical biology

Martin A. Garstens
Office of Naval Research, Washington, D.C.

The central problem of theoretical biology is to supply the missing links between purely descriptive approaches to the field and the powerful mathematical techniques present in modern physics. Traditionally, biology has been a descriptive science. It is the virtue of description that one can keep in view important aspects of the phenomena being observed. With the injection of the successful modern methods of atomic physics into biology, a necessary adjunct has been added, making possible a clearer perception of the underlying mechanisms in living systems. But, with this addition, a restriction in vision and outlook has occurred also. It becomes difficult to see how complete absorption in the atomic details of living mechanisms can ever supply any real insight into the nature and origin of life and its relationship to the rest of the world.

It is the conviction of this author that the middle ground, which unites the purely descriptive analytical aspects of nature, can be found by examining the foundations of statistical mechanics and observing how this field is a synthesis of two streams of experience. The mechanics part suggests its origin in the analytical features of physics and the statistical part suggests origins in the purely descriptive aspects of nature.

In fact, there are strong trends within the field of statistical mechanics to attempt its reduction to its purely analytical part, as in ergodic theory. In our estimation this is a grave mistake, since it deprives statistical mechanics of its most fruitful direction and source of inspiration, which is to provide a synthesis of the analytical aspects of physics with the purely descriptive aspects of nature. In this sense, statistical mechanics properly developed and directed '*is*' from our point of view the field of theoretical biology.

The attempt to stress the all-importance of existent analytical structures supplied by physics is reductionism. Reductionism makes movement and insight into the deeper problems of biology almost impossible. On the other hand, stress on the purely descriptive features of biology as alone relevant for the understanding of living systems (that is, holism) blocks the road to ultimate perception of the connection among these features. In fact, to perceive connections is the essence of insight and understanding. If life is to be understood, such connections

167

must be elucidated. This is a step in the analytical direction. True progress in understanding life must be based on a fusion of the analytical *and* descriptive aspects of living structures.

The role of statistical mechanics in supplying the connection between the analytical and descriptive features of experience can be seen in its efforts to characterize macroscopic variables. Such variables constitute the descriptive aspect of observations. The definition of macroscopic variables in terms of the underlying microscopic features of experience constitutes the very insight which links analytical physics with observed or descriptive aspects of nature.

Two great lessons can be derived from a generalized statistical mechanics approach. One is that all experience in statistical mechanics dictates that relevant macroscopic variables can seldom, if ever, be found by purely deductive methods. For the same reason, the important variables in biology can never be found by such means. From this, we conclude that to understand the nature of life, to probe deeply into the processes which give rise to this phenomena, one must *seek out* the relevant variables wherever they are. One must explore and move out into new experimental domains where such variables may reveal themselves. To look at and restrict oneself to the *present* structures of physics (and even of biology) and attempt to extrapolate from them the new concepts needed to further one's understanding is impossible. The upshot of these remarks is that vigorous descriptive empiricism must never be neglected since this cuts off the discovery of new variables. It is probable that progressive insight into life will require an endless series of such variables.

However, this is not enough. True understanding requires that the variables be seen in their relationship to each other. This is where the second lesson enters the picture. Having selected or perceived the relevant variables, the connections or laws holding among them must be supplied, as they are within statistical mechanics.

Statistical mechanics seems to be so tightly bound to the analytical procedures of physics that it is not obvious how it could be extended to incorporate the descriptive features of a field like biology. This can be done, we believe, by carrying through an extension both of statistical mechanics towards statistics and vice versa, so that there is a natural common ground between them. Thus, the inductive procedures of statistics which are suitable for descriptive phenomena must be incorporated within the structure of statistical mechanics. Conversely, the ensembles used in statistical mechanics would then be applicable to biological classes which contain fewer members than the infinite homogeneous ensembles of physics [1].

168

Is this conception of biological theory complete? Can it cover all valid existent or future theoretical structures descriptive of living systems? In looking to the future, does it seem as if the structure is deep or wide enough to deal with important properties of living organisms? To answer these questions best we might see whether current attempts at theories designed to cover aspects of theoretical biology fit in with the statistical mechanical interpretation.

What is demanded of an adequate theory is a structure capable of ultimately answering important questions like: What is the mechanism of evolution? What is the nature of mentality? How did life originate? The source of the required depth in the approach we are suggesting resides in the demand for new variables or observations, obtained empirically by probing all aspects of our experience. This rules out the superficiality resulting from limited experience.

A brief glance at current efforts to formulate biological theory shows that their linkage with a statistical mechanical approach is not unreasonable. Such efforts at linkage have occurred within fields as various as information theory, cybernetics, theory of automata, systems analysis, and in the application of computer science to brain modelling. Information theory, through the works of Jaynes [2], and others, has been made an integral part of statistical mechanics. Here the notion of entropy is directly defined in terms of information. In each of the other fields mentioned, it is clear that biological macroscopic variables and models are proposed as representative of biological systems. The statistical mechanical problem which then arises is how to relate these to the underlying more plentiful numbers of microscopic variables.

We must forgo the idea of a theory of biology which presents a unified and finished picture as an answer to the 'great' questions. Even physics shows a lack of such unity [3]. Large areas such as mechanics, thermodynamics, relativity, and quantum theory still lack the connecting links which would unify them into a single totality. This type of pluralism must be expected in biology to a degree never experienced in physics. In fact the descriptive approach to biology or any other field implies total pluralism. Every observation is a unity in itself. But as the connections between disparate events are unravelled, the degree of pluralism diminishes. The virtue of an inductive statistical approach is that it bases itself on this pluralism. It accepts the facts as they are.

The resignation to pluralism in biology, however, limits understanding. The search for unifying ties among seemingly disparate areas of theory is a necessity, if one is to advance the knowledge of living systems and their origins. Systemization is a basic requirement for understanding. The implication here is that there are no formal limits to our possible understanding of life. In particular the complementary principle cannot constitute such a formal limit to understanding. It is a limit on the possible knowledge of concomitant variables in a system but, as in physics, it cannot prescribe barriers to future knowledge. In any case, the complementary principle and the principle of uncertainty are themselves hypothetical in nature. There is no 'absolute' knowledge of anything. To prescribe a formal boundary to our knowledge is a statement of absolute limitation.

Additional insight into the role statistical mechanics can play in biology can be gained by looking at current problems in statistical mechanics and in the related area of solid state physics. They are very suggestive of the kind of problems that need solution in biology and of approaches to solutions.

One of the central problems in non-equilibrium statistical mechanics is that of 'macroscopic causality'. How can the description of the approach to equilibrium in a system which depends on a vast number of molecular variables be contracted to a small number of macroscopic variables in a closed and causal way? By closed, we mean that no additional macroscopic variables are needed to characterize the system, and by causal, we mean that predictions can be made by means of the relationships holding among these variables. It is obvious that this problem looms large and is central in a theoretical formulation of biology.

Closely related to the question of macroscopic causality and also very suggestive of problems facing theoretical biologists is the fact mentioned by the great mathematician, Poincaré, that ignorance of the initial state of a system is a prerequisite for the application of statistical methods to closed and deterministic forms of it. It is found that the exact form of an initially smooth state distribution does not matter in the consequent evolution of the probabilities of the various macroscopic states. May we not assume that important macroscopic variables may occur in complex biological systems which *do* depend on the initial state distribution (as in evolutionary biological processes)? How such variables are to be related to those considered in current statistical mechanics discussion is of crucial importance to biology. Considerations along these lines can be found in the works of G. Ludwig [4], where attempts to define clearly macrovariables in terms of the underlying microvariables assume very great significance.

Solid state physics, like biology, is concerned with the many-body problem. The crucial difficulty in both fields is that the many-body problem cannot be solved in terms of quantum theory alone [5]. In the field of solids (as in statistical mechanics) important advances can be made only by getting around the many-body difficulty. Hypotheses of a general nature must be introduced.

M.A. Garstens

These hypotheses must eliminate the need for solving the many-body problem in general and yet be relevant to the material under investigation. As an example, a very powerful tool in solid state physics is the one-electron theory, based on the Hartree–Foch equations. By assuming that the electrons are statistically independent, the one-electron theory can be used as a many-electron theory. Thus, the wave function for the latter, in the general case

$$\Psi(\pi_1, \pi_2, \ldots \ldots \pi_N)$$

reduces to the much simpler product wave function

$$\Psi_1(\pi_1)\, \Psi_2(\pi_2) \ldots \ldots \Psi_N(\pi_N)$$

under this assumption. This example is typical of many approaches to solid state problems. Similar approaches must be used in biological theory. In doing so, it must be kept in mind that, for closed systems, a lack of knowledge of the initial state is a prerequisite for the application of statistical methods. For open systems, the initial states play a significant role and are important for the development of biological theory. In fact, in biology, initial states play a more critical role than they do in the physics of solids. It is this aspect of statistical mechanical theory which needs amplification if greater application to biology is sought.

A whole spectrum of varying initial states can be envisaged which can distinguish problems in physics from those in biology. If there is a complete lack of knowledge of initial states, a purely statistical theory is suitable. As additional relationships are observed, the strictly statistical structure must be modified to accommodate them. Thus, by combining classical physics with statistics one obtains classical statistical mechanics. Similarly with quantum theory, quantum statistical mechanics results. The notion that one can start from a set of principles such as mechanics or quantum mechanics and deduce the characteristics of the rest of nature (reductionism) is simply not true. Rather, one must always envisage an interplay between statistics or empirical observation (bringing irreducible new facts and generalities into sight) and the body of established, coherently integrated knowledge carried over from the past. To see this more clearly, it would be necessary to examine some of the basic problems still remaining in quantum physics, statistical mechanics, statistics, and probability theory. In this author's estimation, these four areas need to be drawn into a coherent synthesis if biological theory is to be developed. Statistical mechanics, where the beginnings of such a synthesis is discernible, gives encouragement in this belief. Thus, it is now possible to distinguish two aspects of statistical mechanics :

171

Statistical mechanics and theoretical biology

1. *Fundamental* properties of macroscopic systems dealing, for example, with the fundamental theorem of thermodynamics, with the use of stochastic equations to describe the approach to equilibrium, and so on. This aspect attempts to understand the most general properties of macroscopic systems from fundamental mechanical or quantum mechanical equations.

2. *Special* properties dealing with problems like heat conductivity in solids, transport phenomena in gases using Boltzmann's collision equation, and others. Consideration of statistics and probability theory suggests that biological theory must fall under this heading.

An examination of the foundations of modern atomic physics reveals a lack of the type of solidity usually associated with classical physics. Instead, the notions of statistics and probability play an important role, implying the basic principle of indeterminacy. Statistical mechanics, as it developed, was felt to be an applied field, a feeling reinforced by the growth of the ergodic program. The established principles of physics were thought to be sufficient to resolve the many-body problem. But it is the basic contention here that this is erroneous. Solutions of the many-body problem as well as those of statistical mechanics require auxiliary assumptions of importance equal to those of quantum or classical physics. These auxiliary assumptions must apply in the overall theory and be uniquely dependent on the particular system under examination. With the continued growth of statistical mechanics the incursion of statistical methods increases. Under more general circumstances, the mechanical (classical or quantum) content can be expected to diminish to the point where even the need of a Hamiltonian disappears. Here theory approaches most closely the area of statistics itself. Finally, in examining the field of statistics one begins to probe the very foundation of all of one's experience in terms of probability theory and the weighing of evidence.

The statistical mechanical approach, we feel, can provide an adequate framework for generalizations of the depth needed to explore fruitfully the nature of living systems. These generalizations should be capable of incorporating an explanation of the mental aspects of such systems and even explain ultimately how living systems can attain knowledge of their very own physical foundations.

M. A. Garstens

References

1. W. M. Elsasser, *The Physical Foundation of Biology* (Pergamon Press: Oxford 1958).
2. E. T. Jaynes, *Phys. Rev. 106* (1957) 620; *Phys. Rev. 108* (1957) 171.
3. L. Tisza, *Rev. Mod. Physics 35* (1963) 151.
4. G. Ludwig, *Axiomatic Quantum Statistics of Macroscopic Systems in Ergodic Theories* (Academic Press: New York 1961).
5. E. Bright Wilson, Jr. (A. Rich, ed.), *Some Remarks on Quantum Chemistry in Structural Chemistry and Molecular Biology* (W. H. Freeman & Co. 1968).

173

Queries and comments on theoretical biology

Edward H. Kerner
University of Delaware

In the Gibbs Laboratory at Yale hangs a little placard bearing a quote from Szent-Györgi : 'The grass is little different from he who mows it'. This has always seemed to me to speak to a fundamental goal towards which theory should be reaching in biology : a global description of the biochemical network *in vivo* that is, in its essence, universal.

Imagine a microscopist who long ago started watching a single cell, and, by optical absorption measurements, kept a tally of the concentrations of a few of the ubiquitous chemical species (for instance, D P N H), generation after generation, always keeping but one cell in the field of view. The record of concentration versus time would be a kind of 'biochemical noise' which, in the nature of the fluctuations it exhibited, would first of all attest to the material being in the living state; secondly, it would characterize this state at such a level and in such terms that theoretical analysis, based primarily on the law of mass action — elementary chemical kinetics, that is — could and should become possible; thirdly, it might be expected that the time-series represented by the record is in first approximation a *stationary* time series, different long strips of the record not being greatly different from one another. The approximate stationarity is in effect the statement of the universality of the living biochemical complex, for even through evolutionary epochs one may surmise that there would be a certain appreciable uniformity of the record. So the short-sighted chemical microscopist does not see in his order of observation the spectacular evolutionary order, does not know that his record started with a primitive animalcule and ended with perhaps a cellular fragment of a complicated being. For him evolution is a subtle overprint across the record of biochemical fluctuations.

It is easy to see that if the microscopist looks too sharply, with increased resolution and magnification down to the molecular scale, he will ruin his specimens and miss the dynamical flow, turning in effect to molecular architecture instead, where the theoretical terms of reference are quantum–mechanical. So to observe and to theorize are undoubtedly valuable works, being as we know leading enterprises of the day. The importance of this sort of inquiry is certainly very great, aiming on the theoretical side to theorize up from first principles into biological realms. While one may have some reservations about this so-called

174

reductionist approach, it clearly has to be pressed. Theoretical understanding is in part the removal of surprise, and the surprise of life may not after all be so unyielding, just as the surprise of superconductivity and that of superfluidity have turned out in recent decades to be not so unyielding. May not the living state be another fairly direct manifestation at macroscopic levels of the quantum laws? May not Planck's constant indeed be observable biologically (that is, in the distinctively biological implications of the molecular architecture, not in the architecture itself, where it is always observable)?

A significant reservation about the reductionist view has been brought out by Elsasser in suggesting 'biotonic' laws which are compatible with quantum laws but not deducible *in principle* from the latter. Another alternative here has seemed to me that — because the peculiarly biological observation must also be reckoned as a physical observation, and because the organism is itself an uncommonly small observer who is closely coupled to what is observed, and who operates in certain respects with classical sharpness while contending on all sides with quantal disperseness — just possibly the domain of biological observation is pressing upon fundamental physical theory. Not as to correctness but as to completeness. The completeness of quantum dynamics has been challenged to some extent since Einstein in 1935. And now Bohm appears to have shown that there is indeed room for 'hidden variables', which the usual scheme of quantum theory had ruled out as a matter of principle. I do not espouse either biotonic law or the incompleteness of quantal law. No clear set of observations seems thus far to compel either. But, with all credit to the pursuit of the reductionist position, compulsion should not really be necessary for exploring other positions.

If our microscopist, tiring of the long biochemical tally, raises his eyes from his instrument, he should of course soon have to turn to be taxonomist, ecologist, geneticist, . . ., in order to assimilate the wild panorama facing him. His powers of observation on this scale are so embarrassingly good as to overwhelm him seemingly with detail. But here too theory-building goes forward, as for instance with population genetics (Wright, Kimura) and eco-dynamics (Lotka, Volterra). This is phenomenological model-making, theorizing from the top down. The terms of reference are principally biological, and it seems to me just here that the evolving corpus of theoretical biology will come to have its own distinctive style, speaking in its own voice to intrinsically biological issues, and not just in a falsetto harking to the physical and chemical background. It is no accident that the central proposition of the large biological world is called the *theory* of evolution. Particularly within the mathematical elaboration of genetics and of ecology must one in time expect to find the full-blown statement of evolutionary concepts, and that statement will be truly a biological theory. Though some of the formal machinery of theoretical physics (Fokker–Planck equations, Gibbs ensembles) may prove useful, significant departures from traditional theoretical ground must be expected and even sought. In any case the indications for indigenous biological theory coming into its own seem rather clear.

Will the eco-genetic theorizing from above have a smooth join, or any join at all, with the molecular–physical from below? Are the biological coordinates coherent with the physical coordinates? A principal testing ground for the conjunction, I suggest, lies in the record, mentioned earlier, of the fluctuations of the *in vivo* biochemical concentrations. For the theoretical tool here is chemical kinetics, where on the one hand phenomenologically introduced rate constants are not exorbitantly far from interpretation on the basis of molecular interaction theory; and where on the other hand the observation is at the macroscopic (single-cellular) level. Of course the chemical kinetics must be that for an open system, indeed allowing for undamped fluctuations, rather than the ordinary kind where decay into chemical equilibrium takes place; and the largeness of the number of chemical species must be duly accounted for, possibly through Gibbs ensemble methods; additionally, the crudity of mass-action laws, in a context where key concentrations may get very low, would in the end have to be corrected. Withal, the start at least has been made (Chance, Goodwin) and the hum of the biochemical noise been made audible. Apart from the technical problems of the kinetics lies the larger question, whether and how the description, discrete and combinatorial in character, of genetical hardware can issue forth from that of the chemical–kinetic software, whose nature is (at least at the start) that of smooth evolution in time according to differential laws.

Finally, a comment on the whole of the theoretical enterprise. It is nice to think that eventually a 'cosmarchy' of theoretical hierarchies can be won. But I argue that the dream of the philosopher's stone must remain a dream, for the master theory would have to write its own self out, and this would seem, in Pascal's word, simply ineffable.

Concepts and theories of growth, development, differentiation and morphogenesis

C. H. Waddington
University of Edinburgh

It is conventional to say that, of all the major classical problems of biology, the ones we understand least are those in the area of 'developmental biology'. Recently some molecular biologists, flushed with their success in discovering the outlines of the process of the control of gene activities in bacteria, have suggested that the problem of development and differentiation in higher organisms would soon yield to attacks along similar lines. In my opinion, this optimism arises from a failure to form a clear theoretical picture of the situation. Only when the theory is adequate is it possible to ask the right questions. At present, there are many areas of developmental biology in which we cannot attain even that modest goal; but we can come near enough to formulating a satisfying theory to see that the real questions that arise are going to be a good deal more difficult to answer than people who are not well acquainted with the field often assume.

In thinking about developmental biology, we need to consider several different levels of theorizing. Perhaps four levels are sufficient to be going on with.

1. Meta-theories concerned with deciding which topics it is profitable to have theories about.

2. Theory Proper, which attempts to define in general terms the logical structure of the problems selected, and to provide an appropriate language in which they can be discussed.

3. General Hypotheses, which specify the types of mechanism invoked to engender these logical structures.

4. Particular Hypotheses, describing how various elements from this array of possible mechanisms are involved in particular cases.

Let us consider from this point of view the four main classical concepts of developmental biology; namely, growth, development, differentiation, and morphogenesis. All these four processes are often referred to as important if not essential characteristics of living systems. A Theoretical Biology certainly has to deal with the phenomena which lead people to make statements of this sort. However, when one examines these phenomena in the light of our modern understanding, one becomes tempted to believe that none of these classical words

177

(except perhaps the last) remains suitable for use today in statements which aim at a reasonable degree of precision. We will discuss them in turn.

A. GROWTH

Meta theory. It has long been recognized that this is a difficult concept to define. As an illustration of an older critical attitude towards it, I will quote some sentences from my own *Principles of embryology*, written in 1954, just before the dawning of the era of 'molecular biology'. 'In everyday usage the word "growth" is used to mean any type of increase in size. . . . There have been two ways of approaching the problem; one of them is to accept the everyday meaning of the word and to study the increases which take place in whole embryos or their parts; the other has attempted to start from some more precisely defined process of growth, and to set up general norms from which the facts as they appear in the development of particular animals can be deduced as special consequences. . . . If we wish to consider a precisely defined process of growth we shall have to find some way of limiting the concept so that it is confined to the increase in size of something which retains a certain similarity to itself. Size may increase merely by the imbibition of water, or by the laying down of relatively inert materials such as shell, bone, cartilage, and so on, which processes differ in kind from the increase in the amount of living material itself. Various definitions have been offered with the purpose of excluding them from the concept of growth as that is required for a precise theory. Gray (1931) speaks of growth as "essentially concerned with the formation of new living material". Medawar (1941) states that "what results from biological growth is itself typically capable of growing". Weiss (1949) gives a more formal definition; growth is "the increase in that part of the molecular population of an organic system which is synthesized within that system", and he further amplifies this, pointing out that it means "the multiplication of that part of the molecular population capable of further continued reproduction". This puts its finger on the important points; if we are trying to formulate a precise concept of growth we must confine it to the increase in the amount of the system which is capable of growing.'

To the hindsight of the present day it is obvious that none of these authors (including myself) had fully grasped the distinction between the production of material which is active in biological ways and of material capable of further replication. We now see that this distinction is absolutely essential, at least at the molecular level. At that level 'that part of the molecular population capable of further continued reproduction', to use Weiss' phrase, *must* contain the genetic

178

material, and is therefore a much more restricted category than 'that part of the molecular organization of an organic system which is synthesized within that system'. Even if we confined our attention to the first phases of synthesis within the system, that is, the production of ribosomes, polymerases, transfer RNAs, messenger RNAs and proteins, we should find that none of these are in themselves capable of further continued reproduction. Even the whole system of them could only continue anything which might possibly qualify as 'growth' for as long as the lifetime of the messenger RNAs. In their basic dependence on the genetic material, even these primary phases of synthesis have no more claim to provide a sufficient substrate for the concept of growth than does the production of such secondary and relatively inert materials as bones, shells, feathers, pigments, and so on. At the molecular level there seems then to be only two clearly definable notions allied to the general concept of growth, namely (1) the replication of the genetic material and (2) the production of certain specified molecules or categories of molecules (where the categories may be of any kind, for example, the whole gamut of enzymatically active proteins, or the non-genetic internal contents of the cell as a whole, or such particular types as the calcified materials, the extracellular collagen, or the like).

When Gray and Medawar attempted to confine growth to the production of 'new living material' or material 'itself typically capable of growing', they were probably thinking of the cellular rather than the molecular level. Even at that level it is very difficult to attach any precise significance to a word with as wide a field of application as 'growth'. Consider the following cases – first at the level of individual cells:

1. A cell of a plant or photosynthetic alga, which takes in from its surroundings only very small molecules (water, CO_2, mineral salts) and synthesizes within the cell itself all the needed macromolecules;

2. An animal cell which takes in from the tissue fluids or blood-stream ready formed peptides, nucleotides, or even proteins;

3. A cell, such as the oocytes of many insect, which has pumped into it, sometimes through a special funnel-like orifice, the whole cytoplasmic and nuclear content of a series of 'nurse cells'.

Again, at the level of groups of similar cells we have such situations as : (1). The number of cells increases, but the size of individual cells becomes smaller (for example, the cleavage stages of eggs, in which the size of the entire group remains much the same) ; (2). The number of cells remains the same, but each cell gets larger (for example, the terminal stages of development of many insect organs from the pupa to the adult) ; (3). New cells are continually being formed (with or without change in size) but cells are also continually being lost by death and disappearance (for example, the skin). It is inevitable that any single definition of growth at the cellular level is bound to conceal important distinctions between these various types of process. We need at least three concepts:

1. The rate of cell multiplication ; 2. The rate of cell death and disappearance ;

3. The rate of enlargement of volume of individual cells. In many situations it would be necessary to subdivide this last category into a number of more precisely defined processes (for example, of cellular differentiation) before the concept could be utilised.

It is perhaps the next most complex biological level, that of the organ or organ system, which presents the most interesting challenge to the interpretation of the concept of growth. The growth of an organized system containing many kinds of cells, such as, for instance, a bone, often if not always involves an exceedingly complex set of processes. For instance the enlargement of the bone requires not only the deposition of new calcified material at certain places by osteoblasts, but also the removal of bony matter from other regions by osteoclasts, as well as shifts in the vascular system, muscle attachments, and so on. When we consider an organ which is constructed out of a number of different elements, as for instance a skull, each of the cranial bones, jawbones, and so on, is undergoing its own series of depositions and erosions. One might expect the system to be too complex for any sense to be made of it. However, in fact it does appear very often that there are some general rules operating on the whole system right across the board of all the individual pieces. D'Arcy Thompson in his *Growth and Form* was one of the first to emphasize this, which he expressed in terms of transformations of a grid of co-ordinates. His illustrations of this are very well known. I give another, perhaps less hackneyed, example (Figure 1) which shows a similar type of phenomenon in the skulls of dogs, where we are seeing the underlying structure rather than only the external form, and where we are dealing with material which is certainly genetically very closely related. We still have no conception of how this sort of co-ordinative transformation can occur in systems comprising many independent units ; the time is more than ripe for a new attack on it with modern molecular methods. This is certainly a definite and identifiable phenomenon which demands a theory.

▶ *Theory and hypothesis.* The meta-theoretical discussion of concepts has shown that we must, in the first place, develop separate (or at least initially separate) theories at the molecular, cellular, and tissue levels.

180

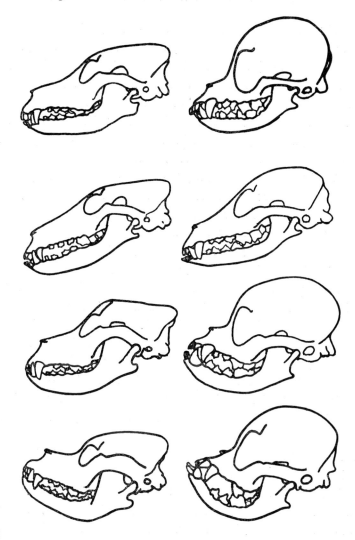

FIGURE 1

▶ 1. *Molecular level.* We need theories and hypotheses about two different categories of molecules : genetic material and non-genetic material which can range all the way from primary proteins to small-molecular deposited materials or 'metabolic sinks', like calcium carbonate, certain pigments, and so on.

   1a. *Genetic material.* We are perhaps just beginning to develop a Theory Proper about the logical structure of the problem. An example at an abstract level of theory is the discussion by Arbib of self-reproducing automata (*Sketches*

181

p. 204) ; at a somewhat more down-to-earth level, there have been considerations such as those of Cairns-Smith and Pattee (*Prolegomena* pp. 57 and 67, and this volume p. 117). At the level of this world's actual biology, an obvious basic factor is the semi-conservative replication of the DNA double helix with all that that involves in the 'unwinding' of coiled strands. But the problem also contains many other questions; about specificities in the beginning and ending of a linearly-operated replication process (which must allow for the epigenetic amplification of particular stretches of DNA, such as the ribosomal cistrons in *Xenopus*, cf. p. 192) ; about some still mysterious tie-up between replication and recombination of strands; and, if the theory is to cover eukaryotes, a replication of the (presumably protein) controllers of gene activity.

We are beginning also to have a few hints about General Hypotheses of the mechanisms involved; DNA polymerase, initiation and termination points which define 'replicons', sigma factors, and the like.

But we are still miles away from good Special Hypotheses. Consider the following cases : 1. How does 'contact inhibition' of culture cells usually bring about cessation of DNA replication? 2. In a germarium of a Drosophila ovary there are sixteen cells, of which fifteen carry on active DNA replication and form highly polytene chromosomes, while the other one becomes the ovum, undergoes meiosis and never becomes more than tetraploid – how? 3. How does infection with a microsporidian parasite provoke enormously increased DNA replication in the polytene chromosomes of salivary gland cells in some insects [1]?

1b. *Non-genetic material*. The task of Theory is to elucidate the shape of the problems involved in controlling the quantities of, on the one hand, primary proteins, and on the other various secondary or tertiary products of biosynthetic pathways. In both areas, Theory seems to be developing, in the main, hand in hand with the elaboration of General Hypotheses. Thus ideas about what is involved in controlling the amounts of primary proteins (for example, the very different numbers of molecules of the proteins coded by the structural genes of the *gal* operon in *E. coli*) arise in connection with hypotheses about the connection between transcription and translation, or re-utilisability of messenger RNAs, and so on. Again, ideas about the control of pools of metabolites seem to be arising mainly in close connection with such General Hypothetical schemes as Stahl's string-processing automata [2], Kacser's studies of enzyme networks [3], and, of course, the oscillatory statistical mechanics of Goodwin, Kerner, Iberall and their school.

▶ 2. *Cellular* (*and organelle*) *level.* As we have seen, meta-theoretical considerations show that different types of cells differ so widely in properties related to increases in cell size and numbers that one cannot hope that general theories applicable to the whole range will be very meaningful. For the functioning of a biological system, it does not seem to be of prime importance whether a living mass is sub-divided into a large number of small cells, or remains as a single large cell, perhaps with a polytene nucleus. The cell is losing its importance as the basic unit of life as attention becomes focused on the more fundamental synthetic processes summarized in the formula 'DNA makes RNA makes protein'. Similar considerations apply to discussions of the increase in size and number of defined intra-cellular organelles, such as mitochondria.

Most growth theories at the cellular level have in fact been developed in connection with special types of cells; in particular those, such as tissue-culture cells or the constitutents of normal epithelia, which engage mainly in proliferation accompanied by little increase in the mass of individual units. In this area there is both a good deal of Theory, dealing for instance with the mathematical investigation of the consequences of assumed rates of multiplication, cell death, and so on, as well as the development of General Hypotheses invoking such mechanisms as specific inhibitors, stimulators and the like [4].

▶ 3. *Tissue level.* Many phenomena which arise, strictly speaking, at the tissue level, for example, the increase in mass of a regenerating liver, can probably be discussed best in terms of the cells composing the tissue; and to that extent fall under the previous heading. The main type of process which seems to call for some treatment at a super-cellular level is that which involves differential growth of the various parts of a system, as illustrated in the examples on p. 181. D'Arcy Thompson's treatment in terms of deformable networks of co-ordinates was a first attempt at a Theoretical statement, and this has been slightly refined by a few authors since (for example, Medawar [5]). Another contribution to the general theory has been the reference to the evidence from genetics that the apparent unitary character of the general pattern must be interpreted as an overall result of a highly complex set of contributory processes (Waddington [6]). But in general we still possess very little in the way of definite Theory in this connection, and almost nothing even of the most general Hypotheses as to the underlying mechanisms which control and co-ordinate the growth of the various parts of such systems. However, we come here very close to the topic of morphogenesis, since a differential growth can be regarded as a dynamic, rather than a merely static, biological form.

Growth, development, differentiation, morphogenesis

DEVELOPMENT AND DIFFERENTIATION

Meta theory.

Development. 'Development' is obviously a portmanteau word, with a very wide application. Probably nobody would be tempted to try to use it in any precise sense. In Darwin's day it was commonly used in senses for which we should now apply the term evolution. During this century development has usually been confined to changes occurring to a single individual during its lifetime. It is a global term, embracing nearly all such changes, whether they are chemical, in the types of cells and their constituent proteins, or morphological. Almost the only limitation on the use of the term is that it normally implies that the changes are progressive and not cyclic, or of a random walk character.

Differentiation. The word obviously refers to a process in which something becomes 'different' in some sense. Like 'development', it was used in the past in the evolutionary context, but it is now almost exclusively used in conjunction with processes within a single life history. However, the development of an embryo involves the arising of two quite different sorts of differences. Many authors, more interested in getting on to describing the particular nugget of experimental fact which they are hoping to add to the edifice of human knowledge than in an intellectual precision, use the word more or less indiscriminately to refer to one or other or both of these processes. A distinction is essential if one wishes to formulate about epigenesis any questions definite enough to be answerable, rather than merely provide some general clue that one is going to talk about some aspect of the development of, for instance, the eye, instead of about its functioning.

One kind of difference arises as time passes. If one looks at one particular region of an egg, it will perhaps at a early stage consist of roughly cuboidal cells, which later become columnar and show the presence of new antigens or other recognizable proteins; then the cells change into a form with a central body and wide-spreading processes, and eventually become a typical part of the central nervous system. This is a progressive sequence of changes of an element in a structure which is always homogeneous at any specified instant. I have suggested the word 'histogenesis' as a name for such time-ordered changes. The word is derived from the root which gives rise to 'histology', but I do not think it need be confined to the tissue level. Differences also arise between spatial parts of a developing system which at the earlier stages appeared to be, and in fact may actually have been, homogeneous in character. A classical case is the upper hemisphere of an amphibian egg. At an early stage no difference of any kind can be discovered between its various regions. At a slightly later stage any part of it which has come in contact with the mesodermal region of the embryo becomes clearly different from parts which have not had this contact, the former developing into the nervous system and the latter into the epidermis. The difference here is the difference between spatially distant regions. These differences, of course, only become expressed in the course of time, but the distinction between the nervous system and the rest of the ectoderm is just as real when they first become operationally distinguishable as it is at a much later stage, when we can detect relatively crude chemical or cell-physiological differences between them. For this sort of differentiation it is space, not time, which is of the essence. For this process I have suggested the word 'regionalisation'.

When one sees a paper headed 'Studies on the Differentiation of the Eye in Species X' one cannot tell whether the author is going to talk about (1) the regionalisation processes by which a spatially specifiable group of cells within the embryo become 'differentiated' from the neighbouring cells which go to form brain or other organs, or (2) how the eye cells change, as time passes, from cells of the early neural-plate type to those of the eye-cup, and finally finish up as cells of the various layers of the retina and the other tissues of the adult eye.

Differentiation is in fact another word that should be dropped from the vocabulary of any decent theoretical biology. It should be replaced by histogenesis or regionalisation, as appropriate.

▶ *Theory and hypothesis.* The meta-theoretical discussion suggested that it would be wise to drop these vague terms, and to formulate Theories and Hypotheses concerning the more precisely defined concepts of histogenesis (change of chemical constitution in time), and regionalisation (subdivision of a unitary system into distinct subsystems); and we shall have to add morphogenesis (change in geometrical form in time), which is more fully discussed later. Although conceptually distinguishable, these three aspects are always closely interrelated in any actual instance of development in a higher organism. The cover-all name for the study of all three aspects is epigenetics.

We have the beginnings of a General Theory of epigenetics as a whole. An epigenetic system contains in the first place a genome, which can be considered as a set of hereditarily transmitted algorithms. They take the place of the 'preformed' elements of older theory. The system will also be characterized by an initial ('pre-formed') regional and morphological structure (for example, that of the ovum at the time of fertilization); but this structure is to be regarded as the result of the operations of genetic algorithms at earlier stages, and is thus

185

ultimately not a primary constituent, as is the genome, but only a secondary one. The Theory goes on to state that these initial conditions define certain possible pathways of change, each pathway being a chreod, that is, having self-stabilising or homeorhetic characteristics which restrain the freedom of variation of the system or its subsystems as they change in time.

From this general theoretical basis, a number of developments must be made to deal with the topics of histogenesis and regionalisation.

In histogenesis, the end-result obviously involves the differential activity of the structural genes coding for the various proteins characteristic of the particular type of cell produced. The main theoretical question is whether we can expect to be satisfied with a theory which involves no more than mechanisms of controlling the activity of individual genes. We have, of course, such a theory, applying in the first instance to the induction of specific enzymes in bacteria, in the well-known repressor–operon theory of Jacob and Monod. The evidence of localised puffing of salivary chromosomes in *Diptera* strongly suggests that processes with at least similar end-results on the activity of single genes (though possibly operating by a different mechanism) occur in the terminal stages of histogenesis in higher organisms. I have argued in many places that inspection of the whole course of a typical histogenetic process in higher organisms suggests that it is necessary for theory to provide for two antecedent stages, of determination and competence, which appear to involve mechanisms of a different kind, and it is now becoming apparent that one should add a third, still earlier, phase, which might be called 'tooling up'.

It has been well known, at least since the time of Spemann (that is, for about half a century) that the critical event which determines the character of the cells into which a given region of an embryo will develop usually occurs some appreciable time before there is any overt evidence of the activity of the genes which code for the proteins characteristic of that cell type. This critical event has therefore been referred to in the embryological literature as 'determination'. A dramatic example of it has recently been described by Hadorn [7]. If the imaginal buds are removed from a larva of Drosophila just before pupation, and injected into the abdomens of adult flies from which the pupation hormones have disappeared, the buds may continue to grow through many cell divisions, and many serial transplantations into new hosts, without showing any sign of producing the histological characteristics of the adult tissues; yet throughout all these cell divisions, each bud retains (with few exceptions) its characteristic 'determination', and the cells derived from an eye bud are ready to produce the proteins characteristic of adult eye-cells, the gonad buds to produce gonadal tissues, and so on, as soon as they are brought in contact with pupation hormones.) There are occasional 'transdeterminations', in which the cells from a bud of one type lose their original characteristics and take on those of some other type, for example, gonadal cells 'transdetermine' into wing cells; but that is another problem which does not concern us at the moment.)

Antecedent to the phase in which overt histogenesis occurs there is not only a phase of determination, but, antecedent to that again, there is a phase in which the determination comes into being. This is referred to as the period of 'competence'. Descriptively, we can say that during a period of competence the system is on a crest separating two or more chreods; during determination it is in some chreod but not necessarily moving along that trajectory ; on activation, it does begin to move. We know rather little about the competent state, even at the relatively unanalytical level of the behaviour of whole cells or cellular aggregates. We do know that when cells are competent to undergo some process of determination, this process can often be brought about by relatively unspecific external agents; thus all the specific constituents required to specify a particular chreod or type of determination must be present already within the cells. We know also that the state of competence may change spontaneously in small isolated groups of cells; a group which responds to a certain external agent at one time by becoming determined as neural tissue may at a later stage of isolation respond to the same agent by becoming lens. But there is not much else we know about competence, and we have a minimum experimental control over it and its alterations.

This logical structure of the theory of histogenesis was developed many years before the discovery of the D N A–R N A–protein story. Probably the most important task for epigenetics at the present time is to discover how to translate the old concepts into the terms of today's (or tomorrow's) molecular biology. At the present time, there are several alternative translations between which we cannot yet decide firmly.

One might be tempted to begin by advancing the extreme view that the whole series of phases – competence, determination, overt histogenesis – does not involve any alterations of the genes on the chromosomes, but that the controls are exerted wholly at the translational or later levels of the synthetic processes. There is good evidence [8] for the presence of 'masked' R N As in embryonic tissues, and for the reality of some form of translation control. There is, however, also evidence that this is not the whole story, since one can actually see alterations occurring at the chromosomes in favourable cases, such as the puffing of bands in polytene chromosomes or the activity of loops in lampbrush chromosomes. In both these cases, we are dealing with activities in highly differentiated cells, that is, with processes which fall within the phase of overt histogenesis. This is the phase in which it would be *a priori* most plausible to suggest that control would be at the translation level. The fact that in these two examples the chromosomes themselves are involved shows that we can certainly not leave control at the gene level ('genotropic control') out of consideration in any of the three phases.

The crucial question in developing an adequate molecular theory of histogenesis is to decide on the interpretation of determination. Once we knew what had happened to the cells in the eye or wing buds of a Drosophila larva to determine them to become those tissues when activated by the pupation hormones, we should be in a better position to understand what this activation might consist of, and what would be required of the competent state which allows the determination to take place. There are at present two fashionable molecular interpretations of determination.

One (recently reviewed, for instance, by Tyler [9]), considers that cells become determined when they start producing the messenger-RNAs for the histologically characteristic proteins of a particular cell type. The fact that the phase of overt histogenesis is not the same as that of determination is accounted for by supposing that these mRNAs are initially produced in a masked form, and need to be unmasked before translation can begin. This theory is based largely on experimental evidence that several histogenetic systems become insensitive to doses of actinomycin-D which suppress new mRNA synthesis at about the same time as they become determined. The weakness of this evidence is, however, that in most of the systems studied the exact time of determination (which is probably not usually a very sudden process anyway) is not very accurately known, so that it is quite possible that there could be an interval after determination is complete and before actinomycin insensitivity develops. In fact, using the Drosophila imaginal bud system, in which there can be a very long period between determination and overt histogenesis, I have shown that these cells do remain sensitive to actinomycin-D well after the completion of their determination, and, indeed, after their overt histogenesis has been triggered off by the pupation hormones also. It therefore seems unsafe to adopt the view that determination is synonymous with the formation of masked mRNAs [10].

The second major view concerns itself mainly with the state of the genes in

188

different tissues, and is usually expressed with less precise reference to the epigenetic processes of determination and histogenesis. It is based on experiments in which there has been extracted from different cell types (for example, liver, pancreas) a material which is considered to be the native DNA–protein complex of the chromosomes; this is then used as a template for the production of RNAs by an *in vitro* system; these RNAs are then studied by hybridizing them with DNA, using various 'competition experiments' in which the RNAs made on the native chromatin templates from different tissues are compared with each other and with RNAs made from deproteinized DNA templates. From such experiments it has been deduced that in the natural chromosomes the genetic DNA is combined with various proteins, which differ characteristically from tissue to tissue, and that it is the presence or absence of a particular type of protein at a given genetic locus which controls the rate at which it is transcribed into mRNA. Different authors have reached somewhat different conclusions as to which type of protein exerts the more important controlling influence, Bonner [11] assigning this role to different categories of histones, while Paul [12] gives greater importance to acidic proteins.

The evidence from which the theory derives has been obtained, as has been said above, mainly from fully differentiated tissues, or from systems which are rapidly growing but can hardly be said to be engaged in typical tissue differentiation (for example, pea seedlings). It could, however, be adapted to the typical process of histogenesis by supposing that determination consists in the formation of appropriate DNA–protein complexes on the chromosomes, in the absence of an adequate transcription mechanism, while overt histogenesis would begin when this mechanism becomes fully functional.

This theory has the merit of focusing attention on the chromosomes, and, as we have seen, the evidence of polytene chromosome puffing and of lampbrush loops shows that the chromosomes must certainly be brought into the picture somewhere. The persistence of the determined state through many cell generations, as in the Drosophila imaginal buds or in many examples in tissue culture, seems also easier to understand if determination involves the genetic materials. But the theory at present still has many weaknesses. The notion that the material extracted from cell nuclei is identical with native chromatin seems difficult to swallow; nor is it obvious that the only substances which can be formed on such templates by the *in vitro* RNA-synthesizing system, and which can affect the hybridization experiments, are in fact natural RNAs (and not, for instance, polymers of simpler types). But the major difficulty is that, even if these

189

substances are all R N As, we know now that in higher organisms it is only the R N As corresponding to the fast-renaturing reiterated stretches of D N A which succeed in forming D N A–R N A hybrids under the conditions of such experiments (Melli and Bishop [13]). Now the majority of these reiterated stretches are not structural genes coding for cytoplasmic proteins – probably none of them are. Thus even if we accept all the premises of the theory it does not follow that these experiments have shown that the chromatin of different cell types is organized (by the proteins attached to its D N As) to produce different m R N As which could control different cytoplasmic syntheses; what has been demonstrated is that the cells are set to produce different types of 'fast R N A' [14] whose function is unknown but is almost certainly not directly related to overt histogenesis.

Neither of these theories is therefore very satisfactory; the first because its basic supposition that determination is synonymous with the production of masked m R N A is probably untrue, the second because it has turned out that what it was dealing with is not m R N A after all. In my opinion, we are in fact not yet in a position to think of formulating a satisfactory theory. There are two crucial pieces of information which we still lack, and which would seem to be essential. The first concerns the mechanisms for controlling whether a gene is inactive as a template, is active as a template for D N A synthesis, or active as a template for R N A synthesis. The problem is by no means completely solved even for the non-proteinaceous chromosomes of prokaryotes; we are right out of sight of understanding the situation when we have to deal with the D N A–protein complexes of eukaryotes. There has recently been much work, particularly by American embryologists, on the question of whether D N A synthesis and massive m R N A synthesis are strictly alternatives or whether they can in some cases be compatible with one another. It seems that in many cases cells stop dividing and synthesizing D N A before they begin intense synthesis of their cell-specific proteins, but in some type of cells (for example, cartilage) the two types of activity can be carried out either simultaneously or alternating over short periods. In any case, whichever way the empirical evidence went, it is by no means clear how it could be interpreted in molecular terms.

To do this we would need, I think, a fuller understanding of a second major problem, namely, the character and in particular the mode of synthesis of the chromosomal proteins. The inheritance of determination through many cell generations strongly suggests that, if the determination depends on the combination of specific controlling proteins with the D N A, the determination brings about the synthesis of the appropriate molecules to allow the condition to be

190

perpetuated. The only way to escape from this would be to suppose that the controlling proteins are in fact fairly unspecific, and are always available in excess, so that the determination would only have to specify which ones were picked up by the new chromosomes at cell division, and not their actual synthesis. However, undoubtedly one of the greatest gaps in our knowledge is any real understanding of where and how, and using what information, the chromosomal proteins are produced.

Such evidence as we have suggests that chromosomal proteins belong in the main to rather simple types, in which one would not expect to find very great complexity and variability in their interaction with the DNA sequences of the genes. Now it is by no means obvious how complex the control of gene activities must be to cause one and the same genome to produce in one instance a liver cell and in the others muscle or nerve cell. Certainly each of these cell types involves the co-ordinated activities of many genes (perhaps several hundred or more); but we have heard in these Symposia authors such as Goodwin and Kauffman explaining how much of this co-ordination might arise within the operations of the synthetic system itself, without having to be specified in detail at the level of the genes. Nevertheless, making all allowances for this spontaneous internal organization, I think one is left with the feeling that the determination of the different cell types in a complex higher organism must demand more complex gene control than most biochemists seem willing to concede as within the capabilities of the rather simple chromosomal proteins alone [140].

However, the amount of information which a system can contain or transmit can be increased by increasing the redundancy; and we have recently learnt of the existence of a great, previously unexpected, redundancy in the genetic system of higher organisms, namely the presence of many families of closely related if not identically reiterated sequences of DNA. So far their function is quite obscure, except in the particular case of the ribosomal cistrons, where the reiteration can plausibly be related to the need to synthesize very large amounts of ribosomal RNA very quickly, particularly in oogenesis. It is tempting to suppose, as a pure speculation, that many of the other reiterated families provide some sort of redundancy mechanism which compensates for the lack of information-carrying capacity of the chromosomal proteins. Perhaps in higher organisms each structural operon has attached to it, not just one or two short controlling sequences as in bacteria, but quite a paragraph of instructions, spelt out by various combinations of reiterated sequences, as a paragraph in English is spelt out with reiterated twenty-six letters of the alphabet [14b].

191

Growth, development, differentiation, morphogenesis

▶ *The 'tooling up period'*. It has always been recognized that important events occur in the development of higher organisms before the onset of the changes, of competence and determination, which lead to overt histogenesis. The general outline of these events has also been known for a long time. There is a period of oogenesis, during which the ovum is formed and 'ripened'; this is followed by fertilization, or some equivalent process of activation; and that again by a period of cleavage, before the first competences for specific tissue chreods can be detected. There seems little need to raise any important objections to this conventional meta theory.

Recent years have seen great advances in our understanding of the Theory Proper of this period. Indeed, we can now go a long way towards formulating satisfactory General Hypotheses about the processes going on in these stages. This is, in fact, the period of development about which our knowledge has been advancing most rapidly. This is partly because during this period the egg has not yet begun to become regionalised, so that there is no difficulty in collecting large homogeneous samples for biochemical analysis, and partly because some of the most important processes involve comparatively large uniform fractions of the genome, that is, some of the reiterated stretches, which again greatly facilitates biochemical work.

The General Theory of this period is that it is largely devoted to the preparation, on a massive scale, of the cellular machinery for protein synthesis. Of this machinery, the ribosomes, particularly the ribosomal RNA, are the easiest for the biochemist to lay hands on. They are produced very massively during oogenesis, by processes which, in the best analysed cases, involve the peculiar step of amplification of the ribosomal DNA cistrons (which are already reiterated) [15]. Something is already known of the provision of nucleases at various stages [16] and there is a considerable literature on the formation of mRNAs in the oocyte, the block to protein synthesis in the ripe ovum, and the effects of fertilization in reactivating this machinery (see Monroy [17] for a recent review). This is one of the few areas of epigenetics where the time has come for theory to take a relatively back seat, at least for a time, with confidence that experimental analysis is sufficiently on the right lines.

▶ *Regionalisation.* The third major meta-theoretical category which we distinguished within the global concept of 'differentiation' was regionalisation. This denotes the arising of chemical differences between parts of a developing system, without any necessary implication that the system itself changes in shape. The gradual separation out from the general cytoplasm of characteristic 'pole plasms' in a spherical molluscan or annelid egg would be an example. Most instances of regionalisation are accompanied, however, by morphogenesis, as for instance when the neural tube becomes regionalised into the fore-, mid- and hind-brains, which swell into their characteristic vesicular shapes; and there are many instances, for example, in the appearance of bristles in definite patterns on parts of an insect's body, in which it is more arbitrary whether one regards the phenomenon as one of the regionalisation or morphogenesis. It is better, therefore, to regard regionalisation and morphogenesis as the two poles of a spectrum of phenomena, most of which partake to some extent of both characteristics.

MORPHOGENESIS

Meta theory. In biological contexts the Greek root 'morph' usually refers to something to do with geometrical shape or form. Strictly speaking the word morphogenesis should mean the coming into being of characteristic and specific form in living organisms. It does not necessarily imply the appearance of any chemical differences between the various parts of the form.

In current practice, the word is rarely used with such precision. For instance, although Needham, in his *Biochemistry and Morphogenesis,* gave it the definition used above, the book itself deals with many aspects of development over and above the coming into being of specific form, and one cannot help feeling that the word is being used somewhat loosely in the title of it. There are many other examples in which morphogenesis is employed as more or less synonymous with development. For instance, C. P. Raven has a book *Morphogenesis: The Analysis of Molluscan Development,* and J. D. Bonner another called *Morphogenesis: An Essay on Development.* Bonner in fact seems to realise that morphogenesis should have a stricter meaning, and he makes a partial concession to this by classifying the constructive processes of development into three types; growth, morphogenetic movements, and differentiation. In this sense, 'morphogenetic movement is the migration of protoplasm which gives rise to changes in form'. He therefore seems to attempt to use the phrase 'morphogenetic movement' as synonomous with the meaning given to the word morphogenesis by Needham. This is, however, not very satisfactory, since in common usage the phrase 'morphogenetic movement' is used to refer to those particular types of morphogenetic processes which involve the extensive migration of regions of the embryo from one position to the other, as for instance in amphibian gastrulation or the dispersion of the cells of the neural crest across the two flanks of the embryo. (Bonner's definitions of growth and differentiation are also rather inadequate. He says 'growth will be used here in the sense of an increase in living matter', but he does not deal with the difficulties, discussed in the first section above, of deciding what is to be counted as 'living matter' in this context. Also he writes '"differentiation" is an increase in the detectable differences in chemical composition . . . of parts of an organism, which occurs between one time during development and another time', so that he has inextricably compounded together the differentiation in time and differentiation in space.

Another complication in the use of the word morphogenesis is referred to by René Thom (*Prolegomena* p. 152), when he points out that in French the word 'morphogenèse' should strictly be used only in connection with the appearance of new organic forms in the course of evolution. Thom, however, goes on to accept the Anglo-Saxon use of morphogenesis in connection with embryonic development. (Thom refers to what he considers an opposition which some authors have set up between morphogenesis and pattern formation. I think, however, that this is a red herring. Thom may well have been referring to my own book *New Patterns in Genetics and Development*, in which after three chapters with the word morphogenesis in their title I have one entitled Biological Patterns, but I carefully say 'there is, of course, no absolute distinction between structures which I choose to refer to as patterns and those which have been considered in the earlier part of this book. . . . Some of these cases are good examples of certain general principles in morphogenesis', which should make it clear that I regard pattern formation as only one aspect of the more general category of morphogenesis. Thom in fact accepts the employment of the term morphogenesis in a more general sense to refer to all processes which create or destroy forms. This is in fact the sense which Needham and I have given to it.

If it is not too late I should like to see the term morphogenesis rescued from those who try to use it as synonomous with development and restricted in its meaning to the sense given in Needham's definition. We certainly need a word to refer to the processes which bring about the orderly arrangement of living systems in space. If morphogenesis has been irretrievably debased then we shall have to invent another word for this meaning, perhaps 'form-generation'.

It is quite difficult to discover in developmental biology phenomena which exhibit nothing but the coming into being of form unadulterated by any other processes : possibly the aggregation of slime mould or other disassociated cells, or again the appearance of a three-rayed instead of five-rayed pattern of protuberances in *Microsterias* would be examples. Much more usually the appearance of new forms is accompanied by changes in the nature of the elements out of

194

which the form is built. In some instances these changes amount to no more than differentiation in time, that is, histogenesis in the sense defined above. An example would be the morphogenetic change which turns the flat neural plate of an amphibian embryo into the cylindrical neural tube. This is accompanied by changes in the shape of the constitutive cells, and these changes in shape are consequences of alterations in chemical composition, but all the cells continued to be neural cells, so that the form changes are accompanied by histogenesis, but not until a later stage by regionalisation. In many other instances morpho-genesis is accompanied not only by histogenesis, but by regionalisation as well, so that different parts of the total form acquire different characteristic properties. An example is the morphogenesis of the mesodermal system, which is going on at much the same time as the rolling up of the neural plate into a neural tube. What was originally a single sheet of mesoderm becomes moulded into a form consisting of a continuous central strand, with a row of more or less cuboidal blocks of tissue lying on each side of it, the blocks merging laterally into a still undivided sheet of material further out to the two sides. Now not only do the cells change their nature in time during this process (histogenesis), but the cells in the different regions change in different directions (regionalisation); the central strand becomes the notochord, the cuboidal blocks the somites, and the lateral undivided sheets the lateral mesoderm.

The most commonly encountered phenomena in development are of this com-plicated type, involving morphogenesis, histogenesis, and regionalisation, all going on simultaneously and probably all to some extent mutually dependent. To refer to the coming into being of the form of the amphibian mesoderm as morphogenesis would obscure the fact that histogenesis and regionalisation are normally also involved, and may be, for all we know, essential for the completion of the purely morphogenetic form changes. I have proposed the word 'indi-viduation' to refer to developmental events in which morphogenesis, histogenesis, and regionalisation are all proceeding simultaneously.

It may be as well to mention that although many examples of histogenesis, regionalisation, morphogenesis, and individuation are accompanied by some increase in the mass of the system which could be referred to as growth, there are also many examples in which no such increase takes place (for instance the development of an early embryo before it begins absorbing nutrients from out-side itself). Thus growth may accompany, but is not an essential part of the other phenomena.

It may clarify matters to consider the types of space that would be required

195

to describe these phenomena. To describe a process of histogenesis we need no more than a space involving a time dimension and dimensions for the concentrations of each of the relevant chemical species; there need be no axes corresponding to the dimensions of real space. The topology of the system would involve a vector field, organized round an attractor which defines a histogenetic chreod. To describe regionalisation we need, in addition, firstly the three dimensions of real space, and secondly a topology involving several chreods separated by catastrophes; but the vector fields defining these chreods operate only on the chemical–temporal co-ordinates of the system, and do not alter the spatial co-ordinates of any element in it. For morphogenesis in its strict sense we need only the four dimensions of real space–time together with a field of vectors which operate to change the space co-ordinates of the elements. For individuation we need the whole battery; spatial, temporal, and chemical axes, and vector fields (with catastrophes) which change both spatial and chemical co-ordinates of the elements.

▶ *Theory and hypothesis.* I have discussed the Theory Proper of morphogenesis and regionalisation at some length in a book published not too many years ago [18] and shall not even attempt to summarize all the arguments here. Suffice to say that I came to the conclusion that there cannot be any one general theory of morphogenesis, at a level more particular than the very abstract ideas of chreods and the like. Biological forms can be produced in many different ways; by the assembly of units which can cohere in certain definite ways only, by copying, or false-copying, on templates; by complexification of systems which originally have a simple disposition in space, and probably several others. Each of these types of process requires its own body of theory. In these meetings we have had contributions by Gmitro and Scriven, Wolpert, Cohen, Waddington, which are all aspects of one or other of the many Theories of Morphogenesis.

Notes and References

1. M. Diaz and C. Pavan, Changes in chromosomes induced by micro-organism infection. *Proc. Natl. Acad. Sc. Wash. 54* (1965) 1321.

2. W. E. Stahl, A computer model of cellular reproduction. *J. Theoret. Biol. 14* (1967) 187.

3. H. Kacser, The kinetic structure of organisms, in *Biological Organization* p. 25 (New York: Academic Press 1963).

4. The following are a few references from the large literature, to indicate the kind of work referred to: W. S. Bullough, *Biol. Rev. 37* (1962) 307; J. Kiefer, *J. Theoret. Biol. 18* (1962) 263; C. Newton, *Bull. math. biophys. 27* (1965) 275; F. M. Williams, *J. Theoret. Biol. 15* (1967) 190.

5. P. B. Medawar, Transformation of shape. *Proc. Roy. Soc. B. 137* (1950) 474.

6. C. H. Waddington, Fields and Gradients, in *Major Problems of Developmental Biology* p. 105 (New York: Academic Press 1967).

7. E. Hadorn, Problems of determination and transdetermination, in *Genetic Control of Differentiation 18* (1965) 48 (Brookhaven Symposia in Biology).

8. For recent reviews, see A. B. Spirin, On masked forms of messenger RNA in early embryogenesis and in other differentiating systems. *Current Topics in Developmental Biology* 1 (1968) *1*; P. B. Gross, Biochemistry of Differentiation. *Ann. Rev. Biochem 37* (1968) 631.

9. A. Tyler, *Devel. Biol. Suppl. 1* (1967) 170.

10. C. H. Waddington and E. Robertson, Determination, activation and actinomycin-D insensitivity. *Nature 221* (1969) 953.

11. J. Bonner, *The molecular biology of development* (Oxford University Press 1965).

12. J. Paul and R. S. Gilmour, Organ specific restriction of transcription in mammalian chromatin. *J. Molec. Biol. 34* (1968) 395.

13. M. Melli and J. O. Bishop, Hybridization between rat liver DNA and complementary RNA. *J. Molec. Biol. 40* (1969) 117.

14. C. H. Waddington, Types of high molecular weight in embryos. *Nature 224* (1969) 269.

14a. Recent evidence concerning initiation and termination factors (*sigma* factors and the like, cf. A. A. Travers & R. R. Burgess, *Nature 222* (1969) 537) suggests that chromosomal proteins may be capable of more specific interaction with DNA than previously thought.

14b. *See* R. J. Britten & E. H. Davidson, *Science 165* (1969) 349; and C. H. Waddington, *Science 168* (1969) 639.

15. See M. Birnstiel, The nucleolus in cell metabolism. *Ann. Rev. plant Physiol. 18* (1967) 25; D. D. Brown, The genes for ribosomal RNA and their transcription during amphibian development, in *Current Topics in developmental Biology, 2* (1967) 48; and E. Perkowska, H. Macgregor and M. Birnstiel, Gene amplification in the oocyte nucleus of mutant and wild type *Xenopus laevis. Nature 217* (1968) 649.

16. J. B. Gordon and H. R. Woodlands, The cytoplasmic control of nuclear activity in animal development. *Biol. Rev. 43* (1968) 233; J. B. Gurdon, Nucleic acid synthesis in embryos and its bearing on cell differentiation. *Essays in Biochemistry, 4* (1969) 26.

17. A. Monroy, Fertilization, in *Comprehensive Biochemistry* p. 1 (ed. Florkin and Stotz) (Amsterdam: Elsevier 1967).

18. C. H. Waddington, *New patterns in genetics and development* (Columbia University Press 1962).

Positional information and pattern formation

L. Wolpert

Middlesex Hospital Medical School

Our almost total ignorance of the mechanism of pattern formation in animals is a crucial deficiency in biological theory. Pattern formation is the spatial organization of cellular differentiation and should be distinguished from molecular differentiation, which refers to the changes occurring within a cell with time and is concerned mainly with the control of the synthesis of specific macromolecules, which are characteristic of the cell type : for example, the processes involved in specifying the synthesis of the muscle proteins in a developing muscle cell or chondroitin sulphate by a cartilage cell. Spatial differentiation, or pattern formation, is the process by which the individual cells within a population are specified to undergo a particular molecular differentiation, which results in a characteristic spatial pattern. For example, the differences between the fore-limb and hind-limb of a tetrapod probably do not lie in the processes of molecular differentiation of, say, the muscle or cartilage cells, but rather in the spatial processes which specify which cells will form cartilage or muscle. Pattern formation should also be distinguished from morphogenesis which refers, in the sense of Waddington, [1, 2], to the forces and mechanisms which bring about changes in shape. A broad view of the problem of pattern formation shows that there is, on the one hand, a great deal of information on the genetics of patterns, and the numerous mutants that affect it, while, on the other hand, morphology has provided us with a very detailed description of the patterns themselves. What is lacking totally is an understanding of the means whereby genetic information within cells is translated into spatial patterns of differentiation. One reason for this is that we do not have a mechanism for pattern formation in cellular terms which could provide a link between molecular biology on the one hand and morphology on the other. An attempt along these lines has already been made with respect to morphogenesis, by investigating the cellular forces responsible for changes in shape [3, 4]. We have suggested that much of the early development of the sea urchin embryo could be accounted for in terms of two cellular operators involving pseudopod activity and changes in cell contact. This analysis suggested that such cellular properties could provide the link between gene action and changes in shape, and also showed how relatively few and simple cellular activities could give rise to complex forms. This analysis also showed that the spatial pattern of

198

the cellular activities which provided the cellular forces was crucial in determining the change in shape.

In considering theories of pattern formation, there seem to be two general requirements which any theory which attempts to account for the spatial pattern of cellular differentiation must fulfil. Firstly, it must provide a simple and reliable means of translating genetic information into very specific spatial patterns of differentiation. Secondly, the theory should be a universal one. It seems most unlikely that the mechanisms whereby genetic information is translated into spatial patterns will differ widely either within an organism or between organisms, though undoubtedly some special mechanisms will exist.

One ought to ask the extent to which there are, or will emerge, general or universal principles which are applicable to development in the same way that there appear to be universal rules for genetics, or, of more relevance, for the transcription and translation of the genetic material at the molecular level. It is too often implicit in embryological thinking that each step in development is a unique or special phenomenon with little general significance [cf. 5]. I would like to suggest that such a view is quite misleading and that there is good reason for believing that there are a set of general and universal principles involved in the translation of genetic information into pattern and form. While some would argue that such a view is gratuitous, it can find some support in consideration of the evolutionary process and our present knowledge of developmental mechanisms. From an evolutionary point of view development is the process whereby the phenotype is specified by the genotype. Selection acts on the phenotype but it is the genotype which is evolving. Considering the universality of the genetic code and of genetic processes, it seems hard to believe that some sort of equally general principles are not involved in the 'translation' of genotype into phenotype. Once a mechanism was established, early on, as it must have been, there seems little reason why it should alter with evolution. In spite of our ignorance of the developmental mechanisms, there is nothing to suggest that general principles will not emerge. On the contrary the very concepts of field and gradient in pattern formation suggest basic underlying principles. In the area of morphogenesis, cell motility and cell contact are increasingly emerging as the basic elements in a wide variety of systems [3, 4].

It is important to realise that while considerable attention is being given to the control of molecular differentiation, few if any of the present lines of thought on the control of gene action lead directly to the solution of spatial pattern formation. We have, for example, the very much exploited model of Jacob and Monod

[6] for the control of gene transcription, but even in its various modifications for eukaryotes this is at least one level of organization removed from problems involving intercellular communication. Dealing as it does with intracellular regulatory phenomena, it is not directly relevant to problems where the cellular bases of the phenomena are far from clear. The concepts in this paper are firmly based on the belief that until the cellular basis of multicellular phenomena such as pattern formation is understood, it is not possible to pose the appropriate molecular questions.

Ideas on the mechanism of pattern formation are deeply rooted in the concepts of induction [7], field [1, 8], individuation [1], gradient and dominance [9, 10], that were elaborated in the 1920s and 1930s. There has been almost no major advance on these ideas since then, with the notable exception of studies on the insect epidermis [11–15], particularly Stern's concept of prepattern [16, 17], and applications of the mathematical analysis of Turing [18, 19]. In general, these ideas neither satisfy the above requirements for a theory nor provide a mechanism for pattern formation in cellular terms. I would suggest that this arises largely from the failure to recognize that the problem of pattern formation is essentially the problem of assigning specific states to an ensemble of identical cells, whose initial states are relatively similar, such that the resulting ensemble of states forms a well-defined spatial pattern.

It is probably convenient to distinguish from the outset between the 'mosaic' type of development in which the specification occurs during the growth of the ensemble from a single cell and in which communication within the system is rather local, and the 'regulative' in which specification occurs in an ensemble of cells, and global interactions are highly relevant. It is almost entirely with the latter that this paper is concerned, though it is recognized that a sharp distinction does not necessarily exist [8]. The effective distinguishing feature between mosaic and regulative development is that when a portion of the system is removed, then the mosaic system will largely lack those regions which the removed portions would normally form, whereas in regulative systems a normal pattern would still be formed.

I have formalized the problem of the regulative development of axial patterns, whose pattern is size-invariant, in terms of the French Flag problem [20]. This problem is concerned with the necessary properties and communication between units arranged in a line, each with three possibilities for molecular differentiation — blue, white, and red — such that the system always forms a French Flag irrespective of the number of units or which parts are removed; that is, the left hand third is always blue, the middle third is always white, and the right hand third always red. This abstraction of the problem corresponds quite well with experimental observations on the early development of sea urchin larvae and regeneration of hydroids as well as a large variety of other systems. For example, the proportions of the mesenchyme, endoderm, and ectoderm of the sea urchin embryo remain constant over about an eight-fold size range; a fragment of hydra, one twentieth its volume, can give rise to an almost complete animal. In more general terms, such systems obey what I have called Spiegelman's rule, which may be stated as follows : the amount of material in a developing or regenerating system that can develop into a particular region or part of a pattern is larger than normally does so [21]. This rule emphasizes the problem of assigning the appropriate states to the cells with reference to the system as a whole.

From an analysis [20, 22, 23] of the French Flag problem it became clear that a solution to the problem lay in providing a mechanism by which the position of each unit from each end was specified and that this positional information could determine the molecular differentiation of the unit; or, in more general terms, the specification of a cell's position with respect to certain reference points in the developing system, that is, positional information. From quite a different approach, Stumpf [11, 12] had concluded from her studies in insect epidermis that the behaviour of a cell was determined by its position within a segment.

It will be suggested in this paper that positional information is the main feature which determines the pattern of cellular differentiation, and that the mechanism of position determination is universal. To put it bluntly a cell knows where it is, and this information specifies the nature of its differentiation, which will be largely determined by its genetic constitution. The main points about the concept of positional information which will be expanded on are :

1. There are mechanisms whereby cells in a developing system may have their position specified with respect to one or more points in the system. When cells have their positional information specified with respect to the same set of points, this constitutes a field.

2. Positional information largely determines, with respect to the cell genome and developmental history, the nature of the molecular differentiation that the cell will undergo. The general process whereby positional information leads to a particular cellular activity or molecular differentiation will be termed the interpretation of the positional information. The specification of positional information in general precedes and is independent of molecular differentiation.

3. Polarity may be defined in relation to the points with respect to which a cell's

position is being specified : it is the direction in which positional information is specified or measured.

4. Positional information may be universal, that is, the same mechanisms that specify positional information may be operative in different fields within the same system as well as in quite different systems from different genera or even phyla.

5. The classical cases of pattern regulation, whether in development or in re-generation, that is, the ability of the system to form the pattern when parts are removed or added, and to show size invariance, as illustrated by the French Flag problem, are largely dependent on the ability of the cells to change their positional information in an appropriate manner and to be able to interpret this change.

The concept of positional information will be shown to provide a unifying conceptual framework for a variety of systems including regeneration of hydroids, sea urchin development, and pattern formation in the insect epidermis. It also may provide some insight into problems of size and growth control. Probably the most important aspect is that it focuses attention on aspects of development which have received far too little attention, particularly where the reference points are and how positional information is specifed. It is hoped that it poses questions concerning pattern formation in a new form, for unless the correct questions are asked there is little hope of obtaining the most revealing answers.

Positional information as here defined has features in common with the double gradient theory of Dalcq [24] and the concept of prepattern as proposed by Stern [16] and extended by Kroeger [13, 14]. The prepattern concept is discussed below. It should be emphasized that the idea that a cell's position is important in development is not a new one, but has been explicit and implicit in the writing of various authors at various times (see, for example, Weiss [25]). However the implications have not been developed nor has the idea of specification of position formed the basis of either theories or experiments in pattern formation.

Induction and its related concepts, which have so dominated embryological thinking, have completely obscured the problems of pattern formation by emphasizing the information coming from some other tissue rather than the response in the tissue which gives rise to the pattern. This is illustrated by experiments showing that head ectoderm of one species of amphibians forms the organs appropriate to its own character when transplanted to the head on another species. 'Since the inducing field can cause the development of organs which are quite foreign to it. . . . it is unlikely that very specific substances are involved. . . . the information coming from the underlying tissues in the toad head which causes a newt balancer to appear can scarcely be very great, since the character of the balancer is determined, not by the transmitted information, but by the reaction to it.' [2]. This illustrates the failure of inductive theory to consider the problem of spatial organization. However, quite early on, the need for some such concept was recognized and gave rise to the field concept. The confusion and ambiguity surrounding the field concept has been discussed many times and it remains a vague postulate that there exist morphogenetic fields which can control the shape of structures [2]. Within the framework of positional information a field becomes that region in which all the cells are having their position specified with respect to the same points or, to put it another way, the same coordinate system. The transplant experiments are immediately interpretable in these terms (see below, Figure 4).

Gradient theory is in some ways much closer to the concept of positional information, particularly the double gradient theories of Bovenri, Runnström, Hörstadius [26], and Dalcq [24], and the emphasis given by Child [9] to the importance of axes. However, as Spemann [7] pointed out, the gradient theory of Child failed to provide a mechanism whereby quantitative differences were translated into pattern. The gradient has usually been interpreted as being a gradient in cellular metabolism or, more vaguely, morphogenetic potential, and has been widely used in accounting for reconsitutution of a pattern when a part is removed, such as is seen in the regeneration of hydroids or planaria. An important feature is the suggestion that the dominant region forms at the high point of the gradient, and that this dominant region both inhibits the formation of a similar region, and that the rest of the regions are formed in relation to it. It has usually remained unspecified as to why the dominant region forms at the high point or how it exerts its influence. It is surprising how little theoretical attention has been given to this problem. Rose [27–9] is one of the very few who have put forward a more detailed mechanism. He suggested that there is a gradient in rate of differentiation and that pattern would be generated if there were a hierarchy of self-limiting reactions, together with the spread of restrictive or inhibitory information from one differentiating region to another. He also suggested that this flow of inhibitory information could be polarized. Our analysis of this system has shown that, while it can generate pattern, it is not capable of regulation, which is an essential feature of many pattern-forming systems [20]. In terms of positional information, gradients may reflect the mechanism whereby position is determined, though the manner in which it operates may be rather different, as will be seen below. With positional information much more emphasis is given to boundary conditions.

203

Positional information and pattern formation

These theories not only fail to satisfy our requirements for a universal mechanism for the translation of genetic information into pattern, but also, even for the specific cases, do not provide a mechanism at the cellular level. They rely in large part on the diffusion of inducers and inhibitors, and it is perhaps significant that no naturally occurring inhibitors or inducers have ever been identified [30]. Another feature, for which the theories are inadequate, is the capacity of many systems to regulate and restore the original pattern when a part is removed.

▶ *Polarity and the specification of positional information.* A cell's positional information is specified with respect to one or more reference points. The identification of such reference points is of great importance, and by no means always obvious; two relatively clear-cut examples are in *Hydra*, where one reference point is almost certainly at the hypostome, and in the early development of the sea urchin embryo the animal and vegetal poles are very likely reference points. The specification of positional information with respect to such points can be essentially of two main types. One is a quantitative variation in some factor, such as concentration of a substance, such that it increases or decreases in some well-defined monotonic way with distance from the reference point. It is thus possible to talk about a positional information / distance curve, several examples of which are shown in Figure 1, and the positional information of the *i*th cell from the end is a_i. The distance is measured in cell number. It is not proposed to discuss in detail here mechanisms whereby the positional information / distance relationship may be generated, but a few examples will be given in order to make the concepts more concrete, by considering a uniaxial array of cells.

A variation in concentration of a substance with distance from the end could

---

FIGURE 1
In (a) a line of cells, N cells long, is shown and the arrow shows the polarity of the system. In (b) three examples of positional information / distance curves are illustrated. Curve I is a case in which there is a linear increase in positional information with distance from the end, and this could represent the increasing concentration of a substance or the phase angle difference between two periodic events spreading from the left as suggested by the Goodwin–Cohen model. Curve II shows another type of relation that could arise for example from active transport of a substance from left to right. Curve III shows a decrease in some property with distance. Note that the value of a_i for the three relationships may be both qualitatively and quantitatively different. In (a) there is a line of N cells and there is one bipolar axis aa'. The *i*th cell, measured in the same direction as the polarity, has positional information $a_i a'_{N-i}$. In (c) there is a sheet of cells, N cells long and M cells wide. It is a two-dimensional field, having two bipolar axes, aa' and $\beta\beta'$. Lines of constant positional information for the aa' axis are at right angles to the long axis; for $\beta\beta'$ parallel to the long axis. Thus, for example, $\beta_0 \beta'_M$ is the lower edge of the sheet. (d) These are two views of a spherical sheet of cells having two dimensions: one bipolar axis aa' and one unipolar axis β. The arrow shows the polarity of the aa' axis. Lines of constant positional information with respect to aa' are shown dashed. The solid arcs are the approximate lines of constant positional information with respect to β. Note the radial symmetry of these axes.

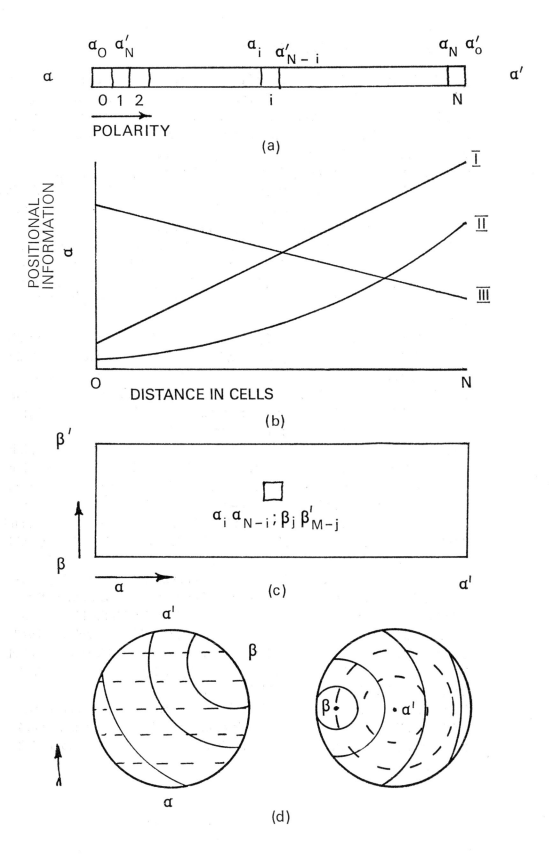

id="1"

(a)

POSITIONAL INFORMATION

α

$\overline{\text{I}}$
$\overline{\text{II}}$
$\overline{\text{III}}$

O

DISTANCE IN CELLS

N

(b)

$\alpha_i \, \alpha_{N-i} \, ; \, \beta_j \, \beta'_{M-j}$

(c)

(d)

Positional information and pattern formation

probably be achieved by some form of active transport. The concentration of the substance would thus specify position with respect to the end. An interesting case is to make the end cells a source and sink respectively of some substance. If the concentration of the substance in these cells were regulated to some fixed value, then there would be a linear gradient between the cells. The absolute value of the substance at any point would specify a cell's position [20].

A particularly interesting, elegant, and important mechanism for the specification of positional information based on the novel principle of wave propagation has been proposed by Goodwin and Cohen [31]. Briefly, they suggest that two periodic signals are propagated from the reference point, the S event and the P event. The P event is propagated from the origin at a definite phase angle difference with respect to the S event, but since it is propagated more slowly the phase angle difference increases with distance from the reference point (Figure 1).

The other type of specification of positional information involves not a quantitative variation in some cellular parameter, but a *qualitative* one : it is essentially a mechanism for cell counting. With such a mechanism the cell at the origin would be A_0, the next cell A_j, and so on to A_N, where $A_1 \ldots _N$ would represent discrete cellular states. These could be represented by membrane states [32]; combinations of different genes; or combinations of different enzymes.

Irrespective of the mechanism whereby positional information is specified, it is clear that it always involves a sense or direction in which it is measured and it is this sense, direction, or ordering relationship, that I here define as the *polarity* of the system. Any system of co-ordinates — and positional information implies a co-ordinate system — requires a direction in which measurement must occur, and this is the polarity. For example, in the examples given above the polarity would determine in which direction the substance was transported in Figure 1. The polarity in the phase shift model is the direction in which the S event is propagated. This in turn may be determined by the frequency of the S event in the cells in the system; under appropriate conditions the cell with the highest frequency will become the pacemaker for the system, and this pacemaker cell will then be the reference point and will also specify the polarity. It is of the greatest importance to recognize that the specification of polarity may be quite distinct from the specification of positional information.

▶ *Rules for the specification of positional information.* Consider N units in an axial array. If the polarity of the system is as indicated by the arrow in Figure 1 the cells will have their positional information specified with respect to the left hand end, since the polarity determines the direction in which the positional information is measured. If we specify this as the α system, then the left hand end cell will be α_0 and each cell will have its position specified with respect to it. The ith cell from the end will have positional information. The following rules may then operate. If a cell is α_i then according to the polarity the cell adjacent to it will become α_{i+1}, in Figure 1 the cell on its right. If a cell does not have its positional information specified, then in some systems, especially those capable of pattern regulation, it will become α_0, and thus the reference point. In some situation if a cell does not have its positional information specified it may not become α_0 but α_m : nevertheless, it will become the reference point for the system. It can be seen that with these rules the polarity effectively defines the ends.

As pointed out above, the physical significance of α_i will depend on the mechanism involved in the generation of positional information. It could represent, for example, the concentration of a substance, the phase difference between two periodic events, a particular state of the membrane, or a particular combination of molecules. At this stage it is preferable to keep the concept in its most general form and define α_i with respect to its generation function, that is, how it varies with distance from α_0.

It is important to realise that positional information of a cell may be specified with respect to a number of points, planes, or surfaces. Positional information may be multi-dimensional. The number of dimensions will be defined by the number of axes. An axis is defined as the line at right angles to surfaces of constant positional information. If there is only one reference point (or surface) the axis will be unipolar. If, however, positional information is specified with respect to both 'ends' then the system is bipolar. For example, in a uniaxial system, whether, for example, it be a single line of cells, or whether they are arranged in a sphere, if there are two reference points one at each end of the axis then (Figure 1) these will be termed the α and α' ends. The positional information on this axis will be referred to as α, α'. It will be assumed that only one polarity need be specified for a bipolar system, the polarity for the α', or β' system will be in the opposite direction.

▶ *Rules for specifying polarity.* Polarity, as defined above, is that ordering relationship which specifies the sense or direction in which positional information is measured. This definition should be compared with that of Rose [28], who considered polarity in terms of flow of information, namely the direction in which inhibitory information moved. The relationship between positional information and polarity is a very close one and a common mechanism may be involved. However at this stage it seems preferable to regard them as separate and to treat the rules for determining polarity as different and distinct from those involved in the generation of positional information once polarity has been established.

A very important insight into how one may consider polarity has come from work on the insect epidermis and particularly from the ideas of Stumpf [11, 12] and Lawrence [15]. Certain structures in the insect epidermis are polarized in that they are orientated in a particular direction, and transplantation experiments can lead to alteration in this polarity. The most important idea to come from such studies is that polarity, as a sense, can be regarded as being determined by the gradient of a substance, a reversal in the direction of the gradient corresponding to a reversal in polarity. Lawrence [15] has suggested a physical analogy in terms of the slope of sand, whose maximum gradient is determined by its angle of friction. In order to account for such gradients in biological terms, he has suggested the possibility of active transport in a direction opposite to that of the gradient. Effectively both Lawrence and Stumpf consider the possibility that the absolute value of the concentration of the substance could provide positional information.

This view of polarity appears capable of explaining a wide variety of phenomena. In order to generalise it and formulate some rules for the determination of polarity I will assume that there is a quantitative measure — the polarity potential — the slope of which determines the polarity. It is only whether the slope is positive or negative that matters, not the value of the gradient, the polarity being in the direction of the slope. Since the polarity determines the reference point in positional information, this will always be at a high point in the polarity potential. In Figure 2a, for example, the polarity potential of a bipolar uniaxial system is shown. The polarity is from left to right and the reference point, a_0, will be at the left hand end at the high point of the polarity potential and a_0' at the low point of the polarity potential. The specification of the reference point at the high point of the polarity potential effectively defines the classic concept of dominance. No further consideration will be given here as to how the polarity potential is maintained, but it is probably, as will be seen, related to positional information. Nevertheless one must not confuse the two concepts. Polarity at a point is a unit vector, positional information a scalar quantity.

The rules for change in polarity will be that there will be a tendency to maintain the same slope and there will be flow from regions of high potential to low potential. (It is quite convenient to bear in mind Lawrence's sand model as a

L. Wolpert

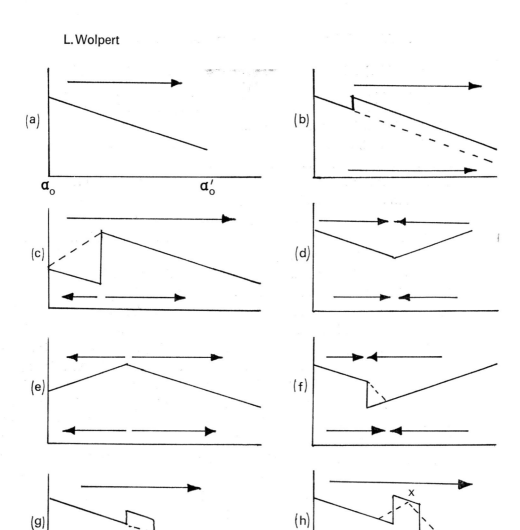

α_0 α_0'

FIGURE 2
Diagrams to illustrate changes in polarity potential with different graft combinations. The ordinate is polarity potential and the abscissa distance. The basic system is illustrated in (a) in which the polarity potential of a bipolar axis is shown. The left hand end is the dominant region. The polarity is determined by the sign of the slope and is shown by an arrow. In (b) to (h) the upper arrow represents the polarity at the time of grafting and the lower arrow the resulting polarities. The dotted line shows changes in polarity potential. Thus in (b) a small increase in polarity potential does not lead to any change in either potential or polarity. However, in (c), a larger change leads to polarity reversal. Other changes are discussed in the text.

209

general guide.) It is also necessary to assume a threshold effect, in that small differences in potential are ironed out without reversal of polarity. These points are illustrated for a variety of cases in Figure 2. In Figure 2b the change in potential is considered to be too small to cause a change in polarity, owing to the threshold effect, whereas a similar graft but with an increased polarity potential, as in Figure 2c does, with time, lead to a reversal of polarity of the left hand end. Thus a graft with the same polarity as the host could result in a reversal of polarity. In Figures 2d and 2e two similar fields are symmetrically joined together with opposite polarities in mirror symmetry. The fields would not be expected to interact and there would be no changes in polarity. However, in Figure 2f, where two fields are opposed with opposite polarities, the higher polarity potential in the left one would lead to a portion of the right field becoming taken into the left one. This would mean that these cells would now have their positional information specified with respect to the left hand end, instead of the right hand end of the system. In Figures 2g and 2h the grafts are not axial ones but involve the insertion of a piece. In Figure 2g the increase in polarity potential is insufficient but in Figure 2h it results in a localized reversal of polarity. This could have very important implication since, in the region of the graft, a new field is established, which is effectively separate from the left position of the orginal field. A new a_0 could be established at the peak (x) of the polarity potential in the graft. One would also expect to find mirror image symmetry about this point.

The interpretation of polarity in terms of polarity potential will be shown to be capable of explaining a wide variety of results. However it is far from being a quantitative theory and two points require immediate comment. The first is that while polarity potential and positional information are treated separately there is little doubt that positional information can affect polarity potential. The second point relates to the whole question of interactions between polarity potentials and the validity of the potential concept. It is, for example, not clear that the potential concept would be valid for the phase shift model of Goodwin and Cohen [31]. In their model, polarity is determined by the pacemaker cell and its ability to entrain adjacent and distant cells, and since, in their model, polarity potential would reflect the frequency, the gradients may well require considerable modification. Nevertheless, the overall picture as shown in Figure 2 might still hold.

▶ *The interpretation of positional information.* It is tempting to use terms like 'translation of positional information' when discussing how the positional information specifies cell behaviour since this can be regarded in a sense as a coding problem, namely, how variation in positional information can specify different cellular activities. However, in order to avoid the terminology associated with DNA–RNA–protein coding, I propose that the term *interpret* be used to describe the process. The overall process whereby positional information specifies a particular cellular state or activity or molecular differentiation will be called the interpretation of the positional information. The mechanisms whereby the positional information is 'read-out' by the cell and changed into a form that leads to the particular activity will be referred to as the *conversion* of positional information. For example, the positional information of cells may be specified by the phase difference between two cyclic processes. This phase difference may be converted into the activation of a specific enzyme, which may in turn be converted into a change in internal ionic concentration, which in turn may lead to activation of a gene coding for a structural protein, which is an enzyme which leads to the formation of a red pigment. One would then say that the cell has interpreted the positional information by forming a red pigment. The interpretation of positional information is, of course, very dependent on the developmental history of the cell and its genome. In fact the terminology allows one to talk about developmental history, or hormones, or mutations, affecting a cell's interpretation of positional information. The concept of conversion then allows one to consider at which stage in the process of interaction conversion is affected.

▶ *The French Flag problem and size invariance.* As an example of the application of the above concepts we can now discuss the French Flag problem. Consider first N cells in a single line. Each cell is capable of molecular differentiation which results in the appearance of blue, white, or red pigment. If the system is unipolar with the polarity as indicated in Figure 3a then the cell at the left end will be a_0 and the ith cell will have positional information a_i. For the system to form a French Flag without size invariance one could have the following rules for interpretation. Between a_0 and a_a, blue ; between a_a and a_{2a}, white ; between a_{2a} and a_{3a}, red. Removal of regions to the right of 3a would have no effect on the pattern if $N > 3a$. However, if part of the Flag is removed from the left end it will regulate provided the remaining N is greater than 3a. For size invariance it is necessary to have a bipolar system [20]. Let the reference point at the left hand end be a and that at the right hand end be a' (Figure 3b). Then each cell will have its position specified with respect to a_0 at the left hand end and a_0' at the right hand end : each cell will have positional information $a_i a'_{N-i}$. This is a bipolar uniaxial field since it has two reference points. In principle, appropriate rules such that the left hand third becomes blue, the middle third white, and the

211

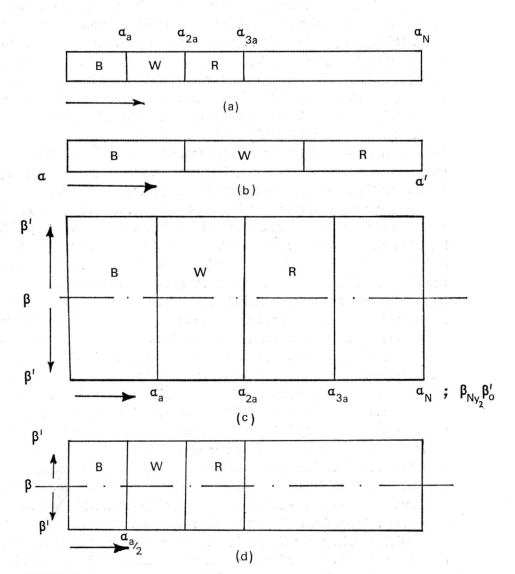

FIGURE 3

Diagrams to illustrate the interpretation of positional information so as to make a French Flag. In (a) a unipolar one-dimensional system, the blue region, for example, is between α_0 and α_a. In (b) the axis is bipolar and the rules for interpretation are such that the whole axis becomes divided up. In (c) a sheet of cells has the α axis unipolar and a bilaterally symmetrical bipolar $\beta\beta'$ axis. Blue is between α_0 and α_a and this should be compared with (d) in which, due to the $\beta\beta'$ axis being shortened, a is reduced to $a/2$ (B, blue ; W, white ; R, red).

212

right hand third red, can always be formulated. These will depend on the nature of the positional information–distance relationship. If, for example, this relationship were a linear one, and identical for both a and a', then the rules for interpretation leading, for example, to blue, could be [22]

$$a_i / a'_{N-i} < \tfrac{1}{2}. \tag{1}$$

This system is capable of considerable regulation and any part could form the French Flag. Consider isolating a portion between a_k and a_l. The positional information at the ends of this piece at the time of isolation will be $a_k\, a'_{N-k}$, $a_l a'_{N-l}$. Since the polarity is from left to right a_k, will become a_0, the dominant region, and a'_{N-l} will become a'_0. There will also be corresponding changes in all the other values of positional information, for example, if there are P cells in the isolated piece, each cell will have positional information $a_i\, a'_{p-i}$, and perfect regulation may occur. This depends, of course, on the cells' ability to change their molecular differentiation when their positional information changes.

The system just described is, from a positional information point of view, bipolar, since the cells are having their position specified with respect to both ends. The description given involved two values, the a and a' ones, and it is important to realise that even in our effectively two-reference-point system only one value for positional information need be specified in order to specify the *relative*, as distinct from the absolute, position of the cell from the ends. For example, in the case described above where one end is a source and the other a sink, the absolute values of the substance at the one-third points is a constant invariant with size [20]. Thus any system which can fix the value of a parameter at both left and right hand ends, and ensures a linear variation between them, would provide a satisfactory solution [cf. 31].

Thus far the problem has only been examined in relation to a one-dimensional system, and it is now necessary to consider the extension to two axes, as would be required if the French Flag were to be formed from a sheet of cells. Consider the unipolar system Figure 3c. Let there be bilateral symmetry such that β_0 is along the midline, and β'_0 at the edge; the $\beta\beta'$ axis is bipolar. Then appropriate rules of interpretation may be specified such that, for example, the value of a which determined a_0 to a_a depends upon the effective sum of β_i and β'_{M-i}: it could be that the greater the sum of $\beta_i\beta'_{M-i}$, the greater a. (Figures 3c and 3d.)

Such rules can provide a solution to the French Flag problem. It is important to realise that with this type of mechanism it is the positional information together with the rules of interpretation that specify the pattern. There is no interaction between the parts of the pattern as such, and it differs in this respect from

213

most other concepts of pattern formation, particularly those that require inductive or inhibitory reactions between differentiated regions. The French Flag problem is possibly misleading in this respect, with its emphasis on discrete regions. Many patterns are perhaps best viewed in terms of a pattern of cellular activity, perhaps every cell having a different molecular differentiation or with a small number of different cellular activities forming the pattern.

It should be clear that the concepts developed here suggest that pattern formation is at least a two-step process. First, and independent of differentiation, is the specification of positional information. Then the cells differentiate according to the interpretation of the positional information. The nature of the interpretation will depend on the cell's genome and its positional history. Clearly such considerations provide the link between pattern formation and the molecular basis of differentiation.

▶ *Positional information and the field concept.* As defined above a field is that region in which all cells are having their positional information specified with respect to the same set of points. The concept of field as classically used is not easy to define and is surrounded by a good deal of controversy. Waddington [1] has pointed out that the term 'field' should be used to refer only to the character of the process occurring in a region or district and should not be used to refer simply to the spatial location of, for example, a presumptive region. It is thus now necessary to show that the definition of field in terms of positional information satisfies the classic requirement for the type of process thought to be taking place.

The term 'field' is used to emphasize the co-ordinated and integrated character of the whole complex of processes. When it is used in connection with the formation of a definite organ with a characteristic individual shape the term can be made more precise by qualifying it as an 'individuation field'. . . . It is clear that this conception of a field implies that the French Flag is an individuation field and the analogy becomes even stronger when some of the operational aspects which define a field are considered. For example [1], if a field is cut in

---

FIGURE 4
Some examples to show some possible implications of the universality of positional information. Consider a rectangular field and two different genotypes. Genotype *fr* results in the interpretation of the positional information so that a French Flag is formed (a) while genotype *sp* results in the cross of St Patrick (b). If, at an early stage, two pieces are interchanged as in (c), and if positional information in the two fields is the same, then the results shown in (d) and (e) will follow : that is, the cells behave according to their genotype and position and are indifferent to the nature of the surrounding tissue. Similarly, if two halves of different genotypes are joined as in (f) a mosaic as in (g) will form (B, blue ; W, white ; R, red).

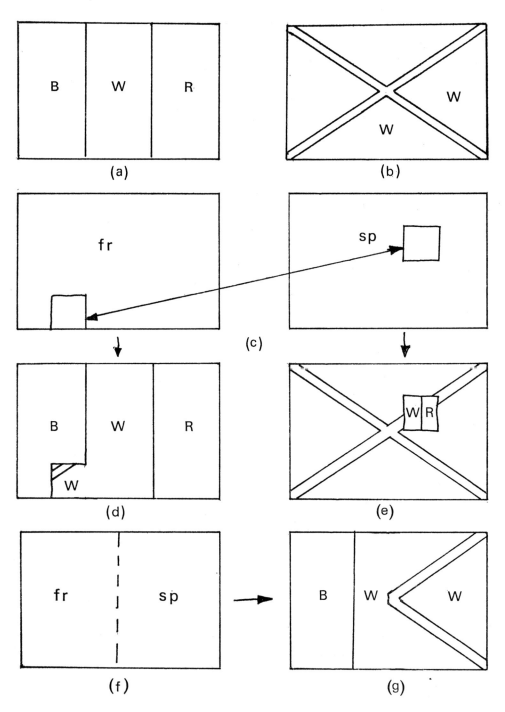

(a)

(b)

(c)

(d)

(e)

(f)

(g)

215

two each half may reconstitute a complete field, so that two whole organs are developed, which is just what the model discussed in Figure 3 is designed to do. Again it is characteristic of a field that if a part, either central or peripheral, is removed, the remainder may compensate for the defect and become complete again, while the isolated part can often become modified into a small but complete field. The concepts of positional information and regulation, together with the specific example of the French Flag problem, can account for all three operational aspects. It is suggested that it is positional information which provides the co-ordinated and integrated character of fields.

It may well be that the mechanism for specifying positional information is a universal one. If this were true then in principle a cell could not distinguish between fields whose geometrical properties or co-ordinate systems were the same. Some operational implications of this are shown in Figure 4. This conception of fields not only provides a firm operational definition for a field and focuses attention on the mechanism involved, but also draws attention for the necessity of identifying the reference points in a field, and considering the specification of boundary values.

The views expressed here should be contrasted with those which invoke specific chemical substances unique to each pattern.

▶ *Positional information and growth control.* As pointed out in the previous section, every cell in a field which has its position specified with respect to two ends of an axis could compute effectively the length of this axis by summing a_i and a'_{N-i}. Thus it is possible to provide rules for interpretation of positional information in relation to cell growth and division such that the length of an axis is controlled in that cell division ceases when the axis is of a given absolute length. The idea that growth control may involve the absolute measurement of one or more lengths in a developing system does not seem to have been considered previously. Most theories of growth control suggest a feed-back system based on the production by the growing cells of some inhibitor such as a chalone [25, 33]. These systems then depend on the dilution out of such inhibition in an external pool; they thus only provide a mechanism for proportionate growth control.

SPECIFIC EXAMPLES OF POSITIONAL INFORMATION AND POLARITY POTENTIAL
Here I will briefly give some specific examples of how the concepts may be applied. For a more detailed analysis see [34].

▶ *Regeneration of hydroids.* Discussion on regeneration of hydroids in terms of

216

L. Wolpert

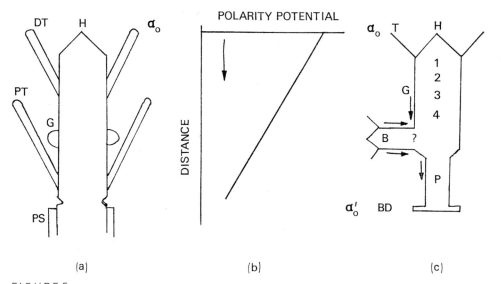

FIGURE 5
(a) Diagram of the hydranth of *Tubularia*. DT, distal tentacles ; H, hypostome ; G, gonophores ; PT, proximal tentacles; PS, *perisarc*. (b) The suggested polarity potential for both *Tubularia* and hydra. (c) Diagram of *hydra*. H, hypostome ; T, tentacles ; BD, basal disc ; B, bud ; GR, gastric region ; P, peduncle. The numbers refer to the regions of the gastric region used in grafting. The question marks show the region at the boundary between the primary and bud axes where the nature or details of the polarity are unknown.

positional information will be confined to observations on hydra and *Tubularia* which are the best studied systems.

Tubularia (Figure 5a) is a colonial hydroid, whose hydranth has remarkable powers of regeneration [28, 35]. This regulation, unlike the French Flag, is not size invariant and *Tubularia*, along its main axis, appears to be a unipolar system.

It will be assumed that the polarity potential in the hydranth decreases disto-proximally, thus the most distal cells will be the reference point for the α axis, which corresponds to the main axis of the hydranth, and will be α_0 as in, for example, Figures 1a and 2a. Then all the cells will have an α_i value depending on the positional information–distance curve, and with appropriate rules of inter-pretation will result in a hydranth : for example, hypostome may form between α_0 and α_a (see Figure 3). If the cell that does not have its positional information specified always becomes α_0, that is α_0 is at the high point of the polarity poten-tial, then normal regeneration will occur if the hydranth, or part of it, is removed. The dominant region is thus at the distal end, as is well known. It is worth emphasizing that, with respect to hydranth form, the rules for interpretation probably involve the specification of cell movements and changes in cell contact

217

at particular a_j and that these give rise to, for example, tentacles. The hydranth may be viewed as comprising a single field and the distinction between, for example, proximal and distal tentacles may be much less than realised since the same cell activities may give rise to both sets of tentacles. This is important since the hydranth is usually viewed as comprising distinct regions, which interact, as in Rose's model [27–9]. For example, in applying his ideas to regeneration of *Tubularia*, Rose regards the regions in a distal proximal direction as comprising hypostome, distal tentacles, gonophores, proximal tentacles, and stolon. It is assumed by him that hypostome must form before distal tentacles, and so on. It is highly questionable, however, as to whether the division of the hydranth into a set of distinct regions, as is essential for his theory of specific inhibition, is in fact legitimate. There is no reason to believe that, from a developmental point of view, they do consititute in fact different and separate regions. The alternative view is to regard them as being the morphological expression of the spatial variation of but a few cellular activities, such as cell adhesion and cell movement [3, 4].

A striking feature of *Tubularia* is the relationship between length, diameter, and pattern. Measurements on large and small hydranths of *Tubularia* have shown in a particularly persuasive manner the proportional decrease in length of the hydranth with decreased diameter [36]. This aspect of pattern formation has been almost entirely neglected although Morgan [37] drew attention to it in relation to the regeneration of both hydroids and planaria. In terms of positional information it implies that the interpretation of a cell of its positional information in the axial direction — a axis — is modified by the positional information measured in the radial direction, the $\beta\beta'$ axis as described above. (See Figures 1 and 3.) For example, the site of distal tentacle formation occurs at increased a values when the diameter is increased, so that effectively the sum of β_i and β'_{M-i} is increased. The failure to recognize this may account for a variety of phenomena not otherwise easily explained, particularly the behaviour of short isolated pieces. The behaviour of short pieces of hydroids have always been rather troublesome for any theory of pattern formation. In general in such short pieces, as shown by Child's studies on *Coryomorpha*, while there is always a disto-proximal ordering of structure, the more proximal structures are often absent. An examination of Child's diagrams [9, Figures 116 and 117] strongly suggests that the scale of organization is related to the radial dimension. In a short piece, if the diameter is relatively large, then, according to the concepts given above, only distal structures will be present. The more normal the length / radius ratio the more complete the scale of organization will be.

218

In contrast to *Tubularia*, hydra seems to be a bipolar system, since the pattern seems size-invariant over quite a range, and so the positional information is specified with respect to the two ends, the hypostome and basal disc. The regions in hydra are less clearly defined, but proceeding disto-proximally there are : hypostome and tentacles ; gastric region ; budding region ; peduncle ; and basal disc (Figure 5c). While it has not been accurately measured, the ratio of the axial length of hydra to its circumference seems about constant in that large hydra have similar proportions to small hydra. This type of regulation would probably involve the α and β axes for positional information, but this important problem will not be considered further here. If we designate the distal and dominant end a_0 and the proximal end a_0' [38] then each cell along the axis will have positional information $a_i a'_{N-i}$ where N is the number of cells along the axis involved in the specification of positional information. Just which cells are, in fact, involved in specifying positional information is, of course, not known. Then regulation will occur as described above with reference to Figure 3. It is also possible to interpret a variety of grafting experiments along lines similar to that for *Tubularia* and in accordance with Figure 2 [34]. The cases illustrated in Figures 2d and 2e behave as would be expected. Of particular interest is the case illustrated in Figure 2c. This effectively predicts that a proximal portion of a gastric region *3* grafted with the same polarity on to a gastric region might have its polarity reversed, tentacles forming at the graft junction and a foot appearing at the distal end. We have recently shown that this occurs in about 30 per cent of the grafts [39].

There is, however, another type of graft which is particularly relevant to hydra, and which demonstrates the 'inducing' properties of the hypostome. This was first demonstrated by Browne [40], who showed that a piece of hypostome grafted into the side of the gastric region will induce a new axis. This is effectively the same as a graft as in Figure 2h which will result in a local reversal of polarity as illustrated in Figure 5d. The new a_0 at X will generate positional information for a new axis. This results in reorientation of material of the host axis and morphogenesis, which is similar to bud morphogenesis [41] but with the important differences that the axis does not detach and is shorter. These phenomena raise very important questions concerning the boundary conditions at the junction between two fields, or two axes, which at present we do not know how to draw. It might be that, at the junction, cells receive the same positional information with respect to the a_0 of both host and induced axis, and that this determines the length of the induced axis : it should increase in length when the junction is further from the hypostome.

219

Positional information and pattern formation

The above interpretations of normal regeneration, and grafting experiments, are very different from those usually given [for example, 27–9, 42, 43] and do not invoke the flow of specific inhibitors or activators between regions. The very concept of regions within the hydranth field may be misleading. There is, for example, not a hypostomal region, but rather a pattern of cellular activities at the distal end determined by positional information. It should be clear that the concepts of polarity potential and positional information provide quite a powerful tool for analysing a variety of phenomena in *Hydra*. It is of particular importance that the phenomenon of induction can be intrepreted largely in terms of polarity changes and dominance in terms of polarity potential.

▶ *Pattern formation in insects and the concept of prepattern*. For me, the most significant contributions to the study of pattern formation over the last thirty years come from the work of Stern [16, 17] on genetic mosaics and the concept of prepattern, together with the experimental work on insects of Kroeger [13,14], Stumpf [11, 12], and Lawrence [15]. This work provides excellent evidence for the concept of positional information and polarity potential, and some evidence for the postulate of universality. As will be seen, the concept of positional information gets over some of the difficulties associated with the concept of prepattern.

The work on genetic mosaics by Stern and his collaborators has shown that there is complete autonomy of cell differentiation in mosaics of different genotypes. His technique essentially provides experiments of the type shown in Figures 4a to 4e. Contiguous areas of different genotypes form their appropriate phenotypes almost regardless of the nature of the neighbouring cells. For example, the sex comb in *Drosophila melanogaster* is located on the first tarsal segment of the fore-leg of males but is absent in females. In genetic mosaics comprising male and female genotypes, the behaviour of cells in the region of the sex comb is autonomous : male cells forming teeth of the sex comb, even if surrounded by female tissue, and female cells being unable to do so even when surrounded by predominantly male tissue. Similar observations indicate autonomy for the engrailed extra sex comb and sexcombless mutants for sex combs, the achaete, theta, and scute mutants for bristle patterns (see [17] for references). These observations have been interpreted in terms of the presence of a prepattern and the competence of cells to respond to singularities in it. This is best explained by reference to sex combs again and quoting Tokunaga and Stern [17] directly — 'The restricted specific area in the male in which a sex comb is formed may be called a regional singularity. An analysis of gynanders (Stern and Hannah, 1950) showed that this singularity exists in both males and females.

220

Its presence evokes a developmental response toward formation of sex comb teeth provided the genotype of the responding cells is male. Female cells lack the competence to form teeth. In so far as the singularity arises during development before formation of the visible pattern of differentiation it constitutes part of a "prepattern" (Stern, 1954, a, b).'

Stern's interpretation of most mutants affecting pattern that have been studied thus far is that the mutant does not lead to a new pattern by changing the prepattern but by changing the competence of the cells to respond to the invariant prepattern. This is an extremely important concept, but has given rise to some difficulties not unlike those associated with the field concept [2, 43, 44]. The immediate problem concerns the nature of the origin of the prepattern. This presents serious difficulties since the prepattern is envisaged by Stern and others as having itself a well-defined pattern, and this problem then becomes how this pattern is specified. For example, the prepattern is viewed by Maynard Smith and Sondhi [19] and Tokunaga and Stern [17] as possibly being represented by variation in the concentration of some substance along an axis, the concentration curve having well-defined singularities such as peaks at specific points. The specification of this pattern seems no more easy than that of the original pattern, though Maynard Smith and Sondhi have attempted to account for the origin of such a prepattern in terms of waves generated by a Turing-like system [18]. This is not very satisfactory, and all the difficulties disappear when the concept of prepattern is interpreted in terms of positional information.

In terms of positional information there is no prepattern in the sense of a pattern with singularities : there is rather the spatial specification of the cells to which the cell's genome can interpret. Cells will behave according to their positional information and genotypes, and more or less independently of their neighbours. In these terms the work relating to prepattern comes within the same conceptual framework as that of sea urchin and limb development, and hydroid regeneration [34].

An important piece of evidence for the universality of positional information comes from the work of Kroeger [13]. Kroeger extended Stern's type of prepattern experiment by manipulative means. Whereas Stern's technique could bring about only mosaics of cells with the same developmental history, Kroeger produced mosaics of regions of the insect with different developmental history. For example, early combined fore-wing and hind-wing imaginal disc of *Ephestia* (grafted together as in Figures 4f and 4g) grew together into a uniform complex. The analysis of the pattern of the hinge parts – sclerites – from such combinations

221

suggested that the prepattern of both fore- and hind-wing were identical. This was suggested by the observation that, however much the sclerites deviated from their normal configuration, the respective parts were always connected in the correct way and tended to form a morphological unit. There was a clear integration between the two fields. Kroeger [13] even went on to speculate that all prepatterns in an insect could be the same. He also emphasized that a distinction should be drawn between the cellular process whereby the prepattern is established and the cellular processes which determine how the cells will respond to the prepattern. These ideas of Kroeger are very similar to, and quite compatible with, those of positional information which I have put forward here. The concept of positional information puts them in a much more general framework.

▶ *Early development of the sea urchin embryo.* The early development of the sea urchin embryo involves its subdivision along its animal–vegetal axis into mesenchyme, endoderm, and ectoderm (Figure 6a). The effect of operative procedures and chemical agents on the relative proportions of these three regions has been widely studied [26, 46]. The classical studies and review of Hörstadius [26], from which all data in this section are drawn unless otherwise stated, have given support to the double gradient concept as a means for controlling the early development. The characteristic feature of these experiments is that, in animalized larvae which result from removal of vegetal material, it is the most vegetal structures that disappear first; and with vegetalized larvae which result from the removal of animal material, it is the most animal structures that are progressively lost. In animalized larvae the ectoderm is proportionately too large; in vegetalized the endoderm. While these and other graftings have firmly established the double gradient concept, involving animal and vegetal gradients, what is completely lacking however is a model of how such gradients are established, regulated, or exert their effect, except in the vaguest of terms. The diagrams given, for example, by Hörstadius [26, Figure 11] imply that the gradients themselves regulate. Needham's diagrams [46] also imply that they change their shape, and that the endoderm–ectoderm border is specified where the two gradients intersect. The interpretation to be proposed here is very different, and will suggest that the gradients themselves do not determine the pattern directly, but serve to specify the values of a_0 and a_0': that is, the value of the positional information at the ends which are the points from which the other cells have their positional information specified.

It should be pointed out first that the early development of the sea urchin shows a size invariance of the main pattern over an eight-fold size range. That is,

L. Wolpert

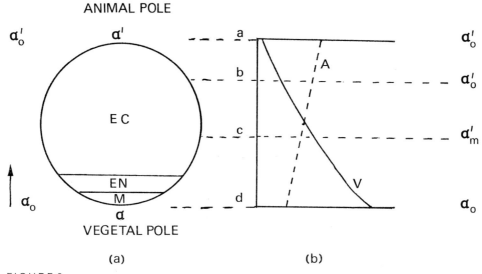

FIGURE 6
(a) Diagram of early sea urchin embryo to show the presumptive regions M, mesenchyme, which normally forms from the micromeres ; EN, endoderm ; EC, ectoderm. a_o is at the vegetal pole and a'_o at the animal pole. (b) shows the possible vegetal, V, and animal, A, gradients. If the region between a and b of the embryo is removed then A/V at the b level is still greater than $K_A = 1$ and the end a' value is a'_o and a normally proportioned embryo results. Removal of the whole animal half, between a and c, results in a vegetalized embryo with disproportionately large endoderm. In this case A/V, at the level of c, is less than 1 and the boundary a' value is a'_m.

blastomeres from the four-cell stage give normal embryos, and, of great importance, so do double volume embryos formed by the fusion of two eggs. It is postulated that positional information is specified with respect to both ends of the animal–vegetal axis, the vegetal pole being a and the animal pole a'. The embryo may then become divided up along this axis into mesenchyme, endoderm, and ectoderm, in a manner which is invariant with size if the rules for interpretation are similar to those for the French Flag as discussed above. What is needed here, in addition, is a mechanism to account for the observed failure of proportionality in animalized and vegetalized larvae.

Consider the early embryo to have two gradients represented by A (animal) and V (vegetal) as in Figure 6. Neither the nature of these gradients nor how they are maintained is known, but they are laid down in the egg. Then it is suggested that the a axis is determined by V, which also determines the polarity potential, the vegetal pole being the dominant region. a' is at the animal pole and is related to A. One then has a system very similar to that in Figure 3 and each cell has its position specified with respect to the aa' axis. If the end cells

223

are a_0 and a_0' then we may expect the systems proportions to be size-invariant. It is proposed that the V/A ratio determines the value of the end or boundary cell for the a axis, and the A/V ratio determines the value of the first or end cell of the a' axis. If $V/A > K_V$, then the first cell will be a_0; if however $V/A < K_V$ then the first cell will be some other value a_m and the less V/A is with respect to K_V, the greater m will be. This means that the positional information distance curve effectively starts m cells along and values between a_0 and a_m are missing (cf. Figure 1). Similar rules are proposed for a': if A/V is greater than K_A, then the first cell is a_0' but if $A/V < K_A$, then it is a_m'.

One can now attempt to interpret briefly some of the more important grafting results in terms of these concepts.

The removal of the micromeres, the most vegetal region, still results in normal development. Doubling the number of micromeres also leads to normal development. This is a somewhat surprising result in terms of classical concepts since the micromeres are the most powerful vegetalizing or organizing region of the embryo. In terms of the new concept, these results merely reflect the fact that the V/A ratio is greater than K_V, both without the micromeres and when their number is doubled. The organizing power of the micromeres is largely due to their being at the high point of the polarity potential, which is at the vegetal end.

Removal of animal material reveals a threshold phenomenon. The behaviour of a vegetal half appears to depend on the position of the third cleavage plane which is at right angles to the animal–vegetal axis and divides it into animal and vegetal halves. In vegetalized larvae which develop from vegetal halves, there is a distinct increase in endoderm at the expense of ectoderm (Figure 6). Some vegetal halves develop normally and Hörstadius [26] suggests that this is due to the level of the third cleavage plane being slightly more towards the animal pole. This would suggest that the critical K_A value for the A/V ratio is about the level of this cleavage plane, and in Figure 6, arbitrarily, has value of 1. If the plane of cleavage passes just above it, that is, on the animal pole side, then $A/V > K_A$ and the vegetal half will be more or less normal since the a' value will start at a_0'. If however, the plane of cleavage is lower down, then $A/V < K_A$ and a' starts at a_m'. This will lead to a decreased ectodermal region and increased endoderm. Effectively, the embryo is behaving as if the regions a_0' to a_m' were present (Figure 6) and a vegetalized embryo forms.

A variety of combinations can lead to the development of normal larvae. One of the most instructive is the effect on the most animal region of adding micromeres. The isolated most-animal region (called an_1) develops into a hollow ball covered with cilia typical of the animal tuft which is the most animal region of the embryo. The same region to which one micromere is added develops into a ball in which the ciliated tuft is initially reduced to a small region. 4 micromeres are required to produce a normal larva. This is in complete accord with the proposed model in which the V / A ratio determines the initial value of a ; 4 micromeres, which have a high V, make the V / A ratio greater than K_V. Fewer micromeres are needed to make an_2 develop normally, since its A value is lower. Practically all the grafting experiments can be interpreted along the above lines : a more quantitative set of conditions and requirements could be obtained relatively easily by simulation studies with a computer.

In these terms the action of vegetalizing and animalizing agents would be to effectively increase the V and A gradients respectively.

This analysis of early sea urchin development differs markedly from the classical ones in that it assigns a very different role to the animal and vegetal gradients to that usually given. Here these gradients do not determine the pattern of cellular differentiation but are concerned with the establishment of the axes and value of the boundary values of positional information at the ends. The organizing properties of the micromeres can find an explanation in terms of polarity changes : their properties in this respect are thus analogous to those of the hypostome of hydra [34].

THE PHYSIOLOGICAL BASIS OF POSITIONAL INFORMATION AND POLARITY POTENTIAL

Very little, if anything, is known about the physiological basis of positional information and polarity potential, largely because appropriate experiments have not been designed with this end in view. For example, an extremely important question is whether or not the specification of positional information, as distinct from its interpretation, involves transcription of the genome or protein synthesis ; and, again, whether cell to cell contact is necessary. One can also ask some general questions to which one may reasonably expect to find some answers in the available literature, or which could be obtained relatively easily by experiment. How quickly is positional information generated ? How far can it be transmitted ? How accurate need positional information be ? — and what is the physical basis of the signals ?

An analysis of these questions in the light of my current knowledge has led me to several surprising conclusions [34]. The first is that the maximum linear size, in terms of cell numbers, of embryonic fields is much smaller than is generally recognized. Most fields, such as the sea urchin blastula, the amphibian neural plate, and the insect segment, are 50 cells long or less. Secondly, while the data is fragmentary, 5 hours seems to be the sort of time required for cells to have their positional information specified. Thirdly, and this is much more tentative, the specification of positional information need not be very precise and one may, in a field 50 cells long, misplace a cell by one or two positions. If these conclusions are true, then what we require is a mechanism for specifying the position of about 30 to 50 cells relatively reliably. By relatively reliably I mean that the whole positional information curve may be shifted along one or two cell lengths, or that its slope may be changed so that a_i becomes a_{i+1}, or even a_{i+2}. What is not permitted is any confusion as to the order of the cells : that is, if the polarity is from left to right then a_i must be always to the left of a_{i+1}. One of the most striking features of pattern formation is the reliability of the serial spatial order of the structures formed. For example, in spite of the developmental flexibility of the amphibian or sea urchin embryo, I know of no experiments which result in the alteration of the serial order of the structures. In general, the problem of precision in development at the cellular level has been grossly neglected. At the tissue level, particularly in relation to the number of modal structures, the problem has been considered by Maynard Smith [47].

If all fields are as small as 50 cells then one must consider how larger structures are formed. This is an extremely important problem which will not be considered here. However two possibilities whereby large numbers of cells can form reliable patterns and yet the size of a field not be greater than 50 cells in any direction are : (1) There may be repeated subdivision of a field into smaller fields. For example, a field 50 cells long may become divided into 5 fields 10 cells long, which in turn grows into fields 50 cells long, and so on. A mechanism of this type has been suggested by Maynard Smith [47] for the number of modal structures. (2) The field may grow and the pattern formed may be determined by cell lineage and local interactions. In mosaic eggs, for example, cellular interactions seem rather local, and the molecular differentiation of a cell is dependent on its lineage. One may thus imagine each of the 50 cells in the line giving rise to, say, 2 daughter cells, each of whose behaviour depends on the polarity of the system and its lineage. Thus cell i will divide into 2 : the left hand one will become i_a and the right hand one i_b. These may divide again to give i_{aa} and i_{ab}; and i_{ba} and i_{bb}, and so on.

Positional information and polarity potential involve intercellular communication, and it will be crucial to determine how this occurs. One question concerns the channel of communication and particularly whether cell to cell contact is necessary [32]. This is particularly important in view of the recent discovery of low ionic resistance junctions between embryonic cells in contact [48]. These so-called functionally coupled junctions would be an extremely attractive candidate for the channel for intercellular communication. It is one of the great virtues of the phase-shift model of Goodwin and Cohen [31] that it makes use of functional coupling and that communication between cells involves the movement of small molecules only between the cells. In their model nothing, in fact, passes physically along the axis; rather it is a wave of activity that is transmitted. Another attractive feature of this model is that polarity potential could have a metabolic basis, since it corresponds to a frequency of oscillation. This would be in line with Child's ideas [9] on the existence of some relationship between polarity and metabolic gradients. Other models could involve the transmission of numerous informational macromolecules between cells and yet others could rely on membrane interaction [32].

What is required is experiments designed with these problems in mind. In general terms one may anticipate that if positional information and polarity potential are universal features of a field system they will make use of very basic cellular properties, and one would not be surprised to find, for example, that they make use of respiratory pathways or cell structures associated with cell division.

CONCLUSIONS

The concepts of positional information and polarity potential seem capable of providing a conceptual framework, within which a wide variety of patterns formed from fields can be discussed and explained at a relatively crude level. A universal mechanism for pattern formation, whereby genetic information is translated into spatial patterns of molecular differentiation, remains a real possibility and a useful working hypothesis. The type of analysis used here requires, surprisingly, quite an intellectual reorientation for those brought up on classical concepts of induction, organizers, and the rather uncritical use of gradient concepts. While the idea that position determines cell behaviour is well known, the implications of this, and the possibility of effectively establishing a co-ordinate system, have not been explored previously. In a sense, the ideas put forward are relatively simple and perhaps do little more than re-describe known phenomena in new terms. Their value lies in providing a general conceptual framework and in defining more clearly the problems involved. While it would be possible to design experiments to test the concepts and invalidate them by, for example, demonstrating an absence of universality, or polarity changes not consistent with the concept of

227

polarity potential, this is probably not the most useful approach. Of far greater importance is the design of experiments and theories to determine how positional information is specified. The development of the phase-shift model of Goodwin and Cohen [31] is the outstanding example of the latter. On the experimental side a variety of questions as to mechanism can be posed, such as, for example, the involvement of the genome in the specification of positional information, and the nature of intercellular communication. That one can discuss the problem in terms of specifying 50 cells relatively reliably is in itself surprising and encouraging. It also becomes necessary, for example, to design experiments to determine the nature of the boundaries between fields and the site of reference points.

It is of interest to note that a universal mechanism for specifying positional information would have implications for evolution. It would effectively provide a co-ordinate system which would enable local changes in a pattern to occur without affecting other regions. One would expect this system to be very stable since any change in the specification of positional information would affect all fields in the organism. The change in the genome of an organism with evolution may in one sense be viewed as providing a mechanism whereby the interpretation of the cells' positional information is altered. In these terms one can perhaps begin to understand the apparent transformation of co-ordinates with evolution of certain structures described by D'Arcy Thompson [49]. It is also not surprising that each cell should contain a great deal of genetic information.

While one is acutely conscious of how ignorant we are of the physiological basis of positional information and polarity, it is hoped that it will have provided a useful unifying framework and will give new meaning to such concepts as gradient, induction, dominance, and field. It is primarily hoped that it will redirect attention to new sorts of approaches and problems, for unless appropriate questions are asked there is little chance of finding out how genetic information is translated into cellular patterns.

This work is supported by the Nuffield Foundation.
I am very grateful for the opportunity afforded by the Serbelloni meeting for the invaluable discussions with Dr Brian Goodwin and Dr Morrell Cohen.

L. Wolpert

References

1. C. H. Waddington, *Principles of Embryology* (Allen and Unwin: London 1954).
2. C. H. Waddington, *New Patterns in Genetics and Development* (Columbia: University Press 1962).
3. T. Gustafson and L. Wolpert, *Int. Rev. Cytol. 15* (1963) 139.
4. T. Gustafson and L. Wolpert, *Biol. Rev. 42* (1967) 442.
5. J. Lederberg, in (A. A. Moscona and A. Monroy eds.) *Current Topics in Developmental Biology I* (Academic Press: New York 1967).
6. F. Jacob and J. Monod, in (M. Locke, ed.) *Cytodifferentiation and Macromolecular Synthesis* p. 30 (Academic Press: New York 1963).
7. H. Spemann, *Embryonic Development & Induction* (Yale University Press 1938).
8. P. Weiss, *Principles of Development* (Hold: New York 1939).
9. C. M. Child, *Patterns and Problems of Development* (Chicago University Press: Chicago 1941).
10. J. S. Huxley and G. R. de Beer, *The Elements of Experimental Embryology* (Cambridge University Press 1934).
11. H. F. Stumpf, *Nature 212* (1966) 430–1.
12. H. F. Stumpf, *Rouse. Arch. EntwMech. Org. 158* (1967) 315.
13. H. Kroeger, Wilhelm. *Roux. Arch. EntwMech. Org. 151* (1959) 113–35.
14. H. Kroeger, *Naturwissenschaften 47* (1960) 148–53.
15. P. A. Lawrence, *J. Exp. Biol. 44* (1966) 607.
16. C. Stern, *Cold Spring Harbor Symp. Quant. Biol. 21* (1956) 375–82.
17. C. Tokunaga and C. Stern, *Develop. Biol. 11* (1965) 50.
18. A. M. Turing, *Phil. Trans. Roy. Soc. Lond. (B) 237* (1952) 37.
19. J. Maynard Smith and K. C. Sondhi, *J. Embryol. Exptl Morphol. 9* (1961) 661.
20. L. Wolpert, in (C. H. Waddington, ed.) *Towards a Theoretical Biology I: Prolegomena* p. 125 (Edinburgh University Press 1968).
21. S. Spiegelman, *Quart Rev. Biol. 20* (1945) 121.
22. M. J. Apter, *Cybernetics and Development* (Pergamon: London 1966).
23. M. Williams (Personal Communication).
24. A. M. Dalcq, *Form and Causality in Early Development* (Cambridge University Press 1938).
25. P. Weiss, in (J. M. Allen, ed.) *The Molecular Control of Cellular Activity* p. 1 (McGraw Hill: New York 1962).
26. S. Hörstadius, *Biol. Rev. 14* (1939) 132.
27. S. M. Rose, *American Naturalist 86* (1952) 337.
28. S. M. Rose, *Biol. Rev. 32* (1957) 351–83.
29. S. M. Rose, Polarized inhibitory control of regional differentiation during regeneration in *Tubularia. Growth 31* (1967) 149–64.
30. C. Grobstein, in (A. V. S. de Reuck and J. Knight, eds.) *Cell Differentiation* p. 131 (Churchill: London 1967).
31. B. Goodwin and M. H. Cohen, *J. Theoret. Biol. 25* (1969) 49. *See also* this vol. p. 231.
32. L. Wolpert and D. Gingell, in (J. Knight & G. E. W. Wolstenholme, ed.) *Ciba Foundation Symposium on Homeostatic Regulators* p. 241 (Churchill: London, 1969).
33. W. S. Bullough, *The Evolution of Differentiation* (Academic Press: London 1967).
34. L. Wolpert, *J. Theoret. Biol. 25* (1969) 1.
35. N. J. Berrill, *Growth, Development and Pattern* (Freeman: San Francisco 1961).
36. N. J. Berril, *J. Exp. Zool. 107* (1948) 455.
37. T. H. Morgan, *Roux. Arch. EntwMech. Org. 10* (1900) 58–119.
38. G. Webster and L. Wolpert, *J. Embryol. Exptl Morphol. 16* (1966) 91.
39. J. Hicklin and L. Wolpert (unpublished.)
40. E. Browne, *J. Exp. Zool. 7* (1909) 1.
41. S. Clarkson and L. Wolpert, *Nature 214* (1967) 780.
42. G. Webster, *J. Embryol. Expti Morphol. 16* (1966) 105.
43. H. Ursprung, in (M. Locke, ed.) *Major Problems in Developmental Biology* p. 177 (Academic Press: New York 1966).
44. C. H. Waddington, in (M. Locke, ed.)

Major Problems in Developmental Biology
p. 105 (Academic Press: New York 1966).
45. T. Gustafson, in (R. Weber, ed.) *The
Biochemistry of Development* p. 139
(Academic Press: New York 1965).
46. J. Needham, *Biochemistry and
Morphogenesis* (Cambridge University Press
1950).

47. J. Maynard Smith, *Proc. Roy. Soc.
(Lond.) B 152* (1960) 397.
48. E. J. Furshpan and D. D. Potter, in
(A. A. Moscona and A. Monroy, eds.)
Current Topics in Developmental Biology
vol. 3 (Academic Press: New York 1968).
49. W. D'Arcy Thompson, *On Growth and
Form* (Cambridge University Press 1961).

A phase-shift model for the spatial and temporal organisation of developing systems

This abstract is reprinted from J. Theoret. Biol. 25 (1969) 49–107.

B. C. Goodwin
University of Sussex
and
M. H. Cohen
University of Chicago

The structure of any differentiated tissue results from a well-defined sequence of events in which the spatial and temporal organization of the developing tissue mass are intimately related. It is as though every cell has access to, and can read, a clock and a map (Wolpert's positional information). A model developed in the present paper is one in which the map arises from wave-like propagation of activity from localized clocks or pacemakers. Individual cells are supposed temporally organized in the sense that biochemical events essential for the control of development recur periodically. This temporal organization of an individual cell is converted by functional coupling between cells into a spatial ordering of the temporal organization. More explicitly a periodic event is postulated which propagates outward from a pacemaker region, synchronizing the tissue and providing a time base for development. Intercellular signalling, entrainment of all cells in the tissue by the fastest cells in the pacemaker region, and a refractory period to guarantee unidirectional propagation are the essential features of the propagation; they permit the derivation of a wave equation and a set of boundary conditions. An underlying gradient of frequency of the event establishes the position of the pacemaker region and the sense of propagation. A second event which propagates more slowly than the first provides positional information in the form of a one-dimensional sequence of surfaces of constant phase difference between the two events. A third event is used to regulate the pattern of phase difference and thus establish size-independent structures. The longest trajectory orthogonal to the surfaces of constant phase difference beginning at the pacemaker region and terminating at the regulating region defines a developmental axis of definite polarity. The model is readily extended to more than one axis, i.e. multi-dimensional positional information. It has a high informational capacity and is readily applied to the discussion of particular developmental phenomena. To illustrate its utility, we discuss development and regeneration in *Hydra*,

231

positional in the early amphibian embryo, and the retinal—neural tectal projection of the amphibian visual system. Specific experiments to test for the existence of the postulated periodic events and their consequences are suggested. Some preliminary experimental results on *Hydra* tending to confirm the model are reported. Possible detailed realizations of the model in terms of biochemical control circuits within the cell, are conjectured and discussed to show that the formal features of the model can be realized by well-recognized biochemical processes.

Comment by C.H.Waddington

I congratulate Wolpert on a bold and fresh attack on this most difficult of biological problems. His idea amounts to the argument that patterns do not arise
from the interaction of neighbouring cells, but that each cell has its position
specified in relation to certain reference points, and then interprets its own positional information independently of its neighbours. What is still left vague, of
course, is *how* this positional information is specified; but Wolpert expresses the
hope, or hunch, that it is by some process which will turn out to be universal.
I should like to mention some phenomena which seem to me relevant, and in
some cases not very easy to assimilate into Wolpert's scheme.

1. Information relevant to pattern formation can be transmitted from one tissue
to another. This was the point of my discussion of the induction of head structures in toad ectoderm placed on to a newt embryo, or *vice versa*, which Wolpert mentions on p. 202. When I said that relatively little information was transmitted from the underlying inducing structures to the ectoderm, what I meant was
that the inducer is certainly not instructing the responding tissue how to develop
the details of a toad's mouth or a newt's balancer; all it was doing was to im-
part the information : 'You are the place appropriate for forming a mouth (or a
balancer)'. But that much information was transmitted. If we interpret the pattern
in terms of positional information, then positional information can be transmitted.
It is not clear to me that this transmission could be by the same mechanism as
that which generates the information in the first place ; which makes the postulate of universality of the mechanism seem less plausible.

2. Stern's studies on 'prepatterns' are, of course, interpretable very easily in
terms of Wolpert's theory, but they do not, in my opinion, provide any noticeable
support for it, since they are interpretable equally easily by any theory which
considers a developmental pattern to be a spatial distribution of processes, and
not of things. In the typical case, Stern dealt with the formation of bristles at
particular places on the body of Drosophila, for example, the four bristles at the
four corners of the scutellum. When he found a gene which had the phenotypic
effect of removing one of these bristles, he jumped to the conclusion that it had
altered the pattern. By studying his mosaic flies, he then showed that, although
the bristle had gone, there was still something at that position which could cause
cells of the normal genotype to produce a bristle. He therefore had to admit that
there is something to do with the sites of bristle formation which the bristle-
removing had *not* altered ; and he called this the 'prepattern'. But it is quite

233

unnecessary to postulate any such mystical entity. If one adopts the process-thinking characteristic of embryologists, instead of the thing-thinking characteristic of geneticists, one can say that this gene fails to alter the pattern, but does affect one of the processes which are distributed in the pattern.

Stern went on to select for study a whole series of genes which fail to alter the patterns but only inhibit the expression of them. But there are other genes in Drosophila which genuinely do change patterns. Genes like *four-jointed*, and *dachs*, and several others, cause the legs to develop with four instead of five tarsal joints, and with completely altered patterns of bristles, sex-combs and other structures. Now it is impossible to suppose that a single mutant gene could alter the read-out instructions by which the cells interpret unaltered positional information so as to produce all these different structures in a new pattern. One is forced to conclude that the mutation has altered the basic pattern. (And when, just recently, Stern [1] eventually came around to investigating one of these genes, he was forced to admit that it did alter the prepattern.) Thus, if we interpret pattern in terms of positional information, we have to allow that positional information can be altered by single gene mutations, and that quite a number of different genes can be effective. Moreover, a gene which can drastically alter the pattern of a leg may have no ascertainable effect on patterns in other parts of the body, such as the thorax. Again, this is not so easy to incorporate into a theory of a universal process of generating positional information, although this might be possible for theories like that of Goodwin and Cohen, which postulate a process in a system which has many independent variables.

3. It has to be remembered that pattern formation can take place within single cells, of Protista such as *Stentor* or *Paramecium* (cf. Tartar, Sonneborn), or lower algae such as *Micrasterias*, which I have discussed at some length [2]. General and would-be universal theories of pattern formation must be formulated in terms sufficiently abstract to accommodate phenomena of this kind; that is to say, they must take as their units cytoplasmic regions, which need not contain their own nucelus and rate as cells. The studies on *Micrasterias* have elicited the facts, relevant to some of the questions Wolpert raises, that : (a) suppression of DNA-dependent RNA synthesis has no effect on the transmission of the pattern to a daughter cell, but does reduce the degree to which the pattern is worked out into its full potential elaboration : (b) it is possible (for example, with a micro UV beam) to suppress the expression of an element in the pattern, without thereby altering the pattern itself (cf. Stern's genes which change the superficial appearance of a pattern without actually altering the pattern itself) : (c) there

234

Comment by C.H.Waddington

exist strains in which the phenotypic pattern is altered; it is not known whether this depends on a truly genetic mutation or on some alteration of an epigenetic template, nor is it known whether it is a change in the basic pattern or only an inhibition of the manifestation of some parts of an unchanged pattern – the point I want to make is that here is a case comparable either to Stern's bristle genes, or to my leg genes, occurring in a single cell system, in which the possibility of specifying the positional information of each of a set of nuclei does not arise : (d) there is very considerable size-invariance in *Micrasterias* (and probably in other single-cell) patterns, though I am not sure just how far this extends.

Thus Wolpert's 'positional information' theory seems to me a novel and challenging idea, which is a real contribution to the theory of morphogenesis; but if it is to be universalised, it must be expressed in extremely generalised terms, which can be applied not only to cells, but to parts of cells, and its basic mechanisms must involve many-variable processes, as do the oscillatory phenomena discussed by Goodwin and Cohen.

References

1. C. Stern and C. Tolkunaga, *Proc. natn. Acad. Sc. USA 57* (1967) 658

2 C. H. Waddington, *Symp. Soc. Devel. Biol. 25* (1967) 105; G. G. Selman, *J. Embryol. exp. Morph. 16* (1966) 489.

The seat of the soul

C. Longuet-Higgins
University of Edinburgh

[Scene : The garden of a timeless Italian villa. A Biologist and a Physicist are strolling on the terrace.]

Physicist : I must say, B, you and your friends have done some most impressive work lately on the molecular mechanism of biological replication. I've been looking out for a new set of problems to give my graduate students, and it seems as if your field is one to which we physicists might well address ourselves. Have you any suggestions about research topics in biology which might be ripe for a rigorous physical approach?

Biologist : It's nice to feel you're interested, P, but before I make any detailed suggestions I shall have to ask you whether you want to go on doing physics or to move into biology.

P : I'm not quite sure I understand your question. But suppose I said that we wanted to go into biophysics, would that help?

B : Yes, it would help a bit; I presume you mean that you'd want to work on physical problems relevant to biology. If so, there are plenty.

P : What sort of problems are you thinking of?

B : Well, we'd like to know more about the transport of ions across membranes, for example. Then there are problems in fluid dynamics, such as the flow of blood through elastic capillaries; problems in quantum mechanics, such as the way in which energy quanta move through aggregates of chlorophyll molecules; there are thermodynamic problems such as the effect of pressure on the partial specific volume of water in cartilage . . .

P (interrupting) : Yes, yes, of course. I do appreciate what you say. Obviously there are a lot of important problems of that sort. But, actually, my students are — if I may say so — an unusually bright lot. I think they might be just a little — how shall I say? — under-motivated to tackle problems in hydrodynamics or classical chemistry. Their really strong suit is quantum mechanics, and we give them an intensive course on many-body theory and Green function techniques before they are allowed to start research. Two of them are at a summer school in Uppsala at the moment, hearing about some fascinating work on proton tunnelling — how it affects mutations, cancer, and all that. And another is spend-ing a few months in Paris, where I'm told there's a most lively school of quantum

236

biology. I'd like them to have the chance of being around when the big break-through comes.

B : Sorry, I'm afraid I've lost you. What do you mean, 'the big breakthrough'?

P : Well, I take it we're agreed that within the next three or four years biology is going to be put on some really firm scientific foundations?

B : Er – well; the firmer the better, of course, but I'm not quite clear what you mean by 'foundations'.

P : Well, take chemistry. Forty years ago it was just a kind of advanced cookery. Then, hey presto, along comes Pauling – no, I mean Mulliken – and now it all makes good sense, on the basis of quantum mechanics.

B : I think I see what you're driving at. But don't forget that a tremendous amount of first-rate chemistry – and I include theoretical chemistry – was done long before 1927, and even now that we have the Schrödinger equation there are still an awful lot of chemical facts that are not understood. Take the chemistry of natural products, for instance. People still do it in much the same way as they did fifty years ago. I have a chemist friend who recently synthesized ATP, and he doesn't understand a word of quantum mechanics.

P : All right, so science doesn't always develop logically. But surely you must admit that if we were really prepared to spend a bit of money on an all-out effort, we should soon manage to pin down the quantum-mechanical basis of biology?

B : I'm not quite sure what you're proposing. I happen to think that we have – or rather you physical scientists have – already pinned down the quantum-mechanical basis of biology, as you put it. But as we might have expected, biology is no less mysterious for all that. You might think molecular biology was an exception, but there the real insights have come not from quantum mechanics but from a combination of bacterial genetics, crystallography, and chemistry, and now that we know the structures of a number of proteins we discover that we're not really as interested in them as we thought we should be. It's the way in which they're organized that really seems to count.

P : How do you mean? Surely if we knew all the structures of all the molecules in a cell, we could in principle work out everything about the cell from quantum-statistical mechanics?

B : Well, for a start, you know as well as I do that when someone says something is possible 'in principle' he really means that it is impossible in practice. But even if it were possible in practice, such a ghastly calculation would dissatisfy both of us. A quantum-mechanical calculation on one particular bacterial cell would be incorrect for every other cell, even of the same species, a point which was made by Elsasser when he stressed the heterogeneity of the material with which the biologist has to deal. We biologists, no less than you physicists, want to be able to make valid *general* statements about our material — statements which hold for all the individuals in a species, and even more widely if possible.

P : If I understand you correctly, you are saying that descriptive biology is not enough; that what biology needs is a more quantitative approach. Could we physicists not help you here with, for example, statistical techniques? Statistics has been highly fruitful in physics, and I am told that it is now regarded as one of the biologists's most useful tools.

B : No, I wasn't really thinking along those lines. Of course I wouldn't deny the importance of statistics in experimental biology, but the recent revolution in molecular biology hasn't owed anything much to statistics. On the contrary, the 'central dogma' of the molecular biologist — about the dominant role of DNA and the subservient role of proteins — is framed in descriptive, not quantitative, terms.

P : But surely you can't deny that biology depends upon physics, and that quantitative prediction is the essence of physical law?

B : No, of course not. But you've just used the word 'depends'. Certainly biology depends on physics, but I happen to think that it's not the same subject. We biologists are trying to find the right language in which to talk about living things. The language of a science must be tailored to its subject matter. To a physicist one caterpillar may look very much like another, and both will seem completely different from a butterfly. But to a biologist the butterfly is the same organism as one of the caterpillars, but a totally different organism from the other caterpillar. One of the jobs of the biologist is to find a rationale for grouping together certain individuals or processes and distinguishing them from others, and such a rationale is not going to be founded on ordinary physical notions such as mass, charge, and so on.

P : You said a few moments ago, talking of proteins and such things, that it's the way they are organized which really counts. How does this assertion tie in with what you have just said about the need to find the right language for talking about living things?

B : Well, I should say that the word 'organization' is a very good example of the kind of word that ought to belong in a biologist's vocabulary. The essential thing, though, is to make sure that when challenged we can explain exactly what such a word means. You might say that this is a philosophical problem, if you regard philosophy as the attempt to clarify concepts.

Christopher Longuet-Higgins

P : All right, so we have to try and define 'organization'. But is it an exclusively biological idea? Don't we physicists use it when we are thinking about perfect crystals, for instance?

B : Do you really? When I hear physicists talking about such things, they usually seem to use the word 'order', not 'organization'.

P : All right, but is there a real difference?

B : Yes, and a very important one in my opinion. There are two quite different meanings of 'organization', an active and a passive one. You could I suppose say that when you tidy your desk you produce a greater degree of organization, or that a sonnet manifests a certain kind of organization. But what you would really mean is that you create order, or that a sonnet displays a certain kind of order. I want to talk about organization in the active sense, meaning the process of writing a sonnet or the manner in which you tidy your desk; I should want to say that order is sometimes (but not always) produced by organization.

P : Very well, let's use the word in that way. But you still haven't given me any idea how you would define it more exactly, or how it relates to biology.

B : I can see it's time for me to show my hand. Let's just have a look at the new biology. What has been the most important biological discovery of recent years? Surely the discovery that the processes of life are directed by programs, in the most professional sense of that word. In nature, as opposed to computing labs, the really distinctive thing about living processes is that they manifest pro-grammed activity, while non-living processes do not. All other distinctions which people have tried to draw between life and non-life have come to grief when confronted with some physical system or other. For example, many physical processes produce complicated ordered patterns such as snowflakes or con-vection cells. Feeding, replication, excretion, homeostasis, all these properties are manifested by fire : it consumes fuel, it multiplies, it produces ashes and it's hard to blow out. But it is not programmed.

P : But surely it is programmed? Don't the laws of physics, in particular the Schrödinger equation, constitute the program for combustion, and for every other physical process, living or non-living?

B : No, I can't agree. If the idea of a program has any meaning at all, it must be possible to distinguish different programs from one another; to say that there is just one program for everything is to make nonsense of the word. That is one thing. Another distinctive feature of a program is its conditional character. All programs of any interest contain instructions of the type 'if (dot, dot, dot) then do such and such, otherwise do something else'. The 'something else' may be

239

to enter a particular subroutine, or to move to a different point in the program, either an earlier point or a later one.

P : I think I see what you are getting at. But surely in any developing physical situation, what happens will depend upon various conditions, won't it? How do you distinguish objectively between living processes and other physical processes which are liable to interference – for example, from outside the system under consideration?

B : You have exposed your flank there, P, I am afraid. That word 'consideration' is a tacit admission that scientific description can never be as 'objective' as our grandfathers supposed. Surely we ought to have learned that lesson by now from quantum mechanics if not from philosophy. But I must resist the temptation to score debating points, and try and answer your question in more detail. If we had had this conversation fifty years ago, and I had asserted that the cell was controlled by a program, it might have been very difficult for me to justify this view in a convincing way. But the extraordinary thing, proved by Watson and Crick, is that one can now point to an actual program tape in the heart of the cell, namely the DNA molecule. And more recently people have actually discovered how the characters on the tape are translated into the twenty-letter alphabet of the amino-acids, the basic building blocks of the proteins.

Computing scientists agree that the idea which made their whole subject possible was that of the stored program. Well, it seems that nature made this discovery about 1,000 million years ago.

P : Are you suggesting, then, that life is just programmed activity, in the computer scientist's sense of 'program'? Because if so you will find yourself driven into saying that a computer is alive – at least when it is executing a program, and that strikes me as mildly crazy. How about that for a debating point?

B : Fair enough. But I wouldn't put it past computing scientists to construct a machine which we would have to treat as if it were alive, whatever our metaphysical objections to doing so. I think I should want to point out, though, that the programmed activity which we find in nature is marked by at least one characteristic which hasn't yet been successfully copied by the engineers. In nature the controlling programs do not merely determine the way in which an organism reacts to its environment. They also control the actual construction of the organism, and its replication, including the replication of the programs themselves. This is very important, because the small alterations which sometimes occur during replication lead to phenotypic variations upon which natural selection can then operate. So life is not merely 'programmed activity' but

240

'self-programmed' activity.

P : 'Self-organization', in other words ?

B : Precisely, but I am trying to spell out the meaning of this term more fully than other people who have used it before.

P : I can see that it could be a very useful one for helping one to think about morphogenesis and possibly about evolution. But I am sure you would agree that biology holds other equally fascinating problems. What about the brain for example ? How could brain function be fitted into your scheme ?

B : I think it fits rather well. Ask yourself, what kinds of thing do we really want to know about the brain ? I suggest that what we would like is a detailed account, among other things, of the 'software'. I mean what a computer scientist would mean : the logic of the master program which sees to it that the user's program is properly translated into machine code, and implemented according to his instructions without lousing up the programs of other users. It seems quite reasonable to regard this as the ultimate objective of psychology. Psychology joins hands with physiology at the point where questions about software raise questions about the hardware by which the behavioural programs are implemented.

P : You know, that idea raises rather an interesting question. You pointed out that in molecular biology the actual program tape was discovered rather late in the day. Do you think that we might hope to find a physical embodiment of the master program which underlies our mental activity ?

B : Heaven knows. But it's quite on the cards that there is a special program in charge of all our various subroutines, which must not conflict with each other if we are to behave in an integrated way. And possibly its instructions reside in quite a small part of the brain.

P : The seat of the soul, in fact ?

[Enter two waiters with tea trays from the villa, and the conversation ends abruptly.]

241

List of Participants

Peter **Buneman** Topologist, now with Christopher Longuet-Higgins in the University of Edinburgh.

J. A. ('Jim') **Burns** Computer scientist, pre-doctoral studies on simulation of enzyme networks carried out in the Department of Genetics, University of Edinburgh.

Morrell **Cohen** Physicist. Committee on Mathematical Biology, University of Chicago.

Jack **Cowan** Mathematical Neurobiology. Head of Department and Professor of Mathematical Biology, University of Chicago.

W. M. **Elsasser** Physicist. University of Maryland.

Alex. **Fraser** Geneticist and Animal Breeder. Made the first computer simulations of evolutionary processes.

Martin **Garstens** Physicist. University of Maryland.

Brian **Goodwin** Theoretical Biologist. University of Sussex, Brighton.

Richard **Gregory** Neuroscientist. University of Edinburgh.

Stuart **Kauffman** Pre-doctoral studies of medicine. Now at the Committee on Mathematical Biology, University of Chicago.

Edward H. **Kerner** Physicist. University of Delaware.

Mrs Suzanne **Langer** Aestetician and Philosopher. Department of Philosophy, Connecticut College. Author of *Mind: An Essay on Human Feeling* and other books.

R. ('Dick') **Levins** Ecologist. Department of Zoology, University of Chicago. Author of *Evolution in Changing Environments*.

R. C. ('Dick') **Lewontin** Evolutionary Biologist. Department of Zoology, University of Chicago.

C. **Longuet-Higgins** Professor of Machine Intelligence, University of Edinburgh. Fellow of the Royal Society.

John Maynard **Smith** Geneticist. University of Sussex, Brighton.

243

List of Participants

Howard H. **Pattee**	Physicist. Stanford University.
Ruth **Sager**	Molecular Genetics, specializing in non-chromosomal heredity. Department of Genetics, Hunter College, New York.
René **Thom**	Topologist. Institut des Hautes-Études, Bures-sur-Yvette, Seine-et-Oise, France.
C. H. ('Wad') **Waddington**	Biologist. University of Edinburgh.
Lewis **Wolpert**	Developmental biologist. Department of Biology, Middlesex Hospital Medical School, London.

Author Index

References to extended treatments are given in bold type

245

Author Index

Author Index

Subject Index

248

Subject Index

Subject Index

Subject Index

Subject Index

Subject Index

253